LIFE CHANGING

How Humans are Altering Life on Earth

Helen Pilcher

Illustrated by Amy Agoston

BLOOMSBURY SIGMA
LONDON • OXFORD • NEW YORK • NEW DELHI • SYDNEY

BLOOMSBURY SIGMA
Bloomsbury Publishing Plc
50 Bedford Square, London, WC1B 3DP, UK
29 Earlsfort Terrace, Dublin 2, Ireland

BLOOMSBURY, BLOOMSBURY SIGMA and the Bloomsbury Sigma logo are trademarks of Bloomsbury Publishing Plc

First published in the United Kingdom in 2020

This edition printed in 2021

Photo credits (t = top, b = bottom, l = left, r = right, c = centre)
Colour section: P. 1: © Molly Merrow (t); © Helen Pilcher (b). P. 2: © Helen Pilcher (t); © Richard Pell, Centre of PostNatural History (b). P. 3: © Andrew Digby (t); © Edwin Winkel (b). P. 4: © Auscape / Getty Images. P. 5: © De Agostini Picture Library / Getty Images (t); www.glofish.com (b). P. 6: © Charlie Burrell (t, bl); © Neil Hulme (cl). P. 7: © Larry Wadsworth, College of Veterinary Medicine & Biomedical Sciences, Texas A&M University (t); © Alejo Menchacha, Instituto de Reproducción Animal Uruguay (b). P. 8: © Richard Pell, Centre for PostNatural History (t); © Helen Pilcher (cl, bl, br).

A catalogue record for this book is available from the British Library

Library of Congress Cataloguing-in-Publication data has been applied for

ISBN: PB: 978-1-4729-5672-9

2 4 6 8 10 9 7 5 3

Typeset by Deanta Global Publishing Services, Chennai, India
Printed and bound by CPI Group (UK) Ltd, Croydon CR0 4YY

Bloomsbury Sigma, Book Fifty

For Amy, Jess, Sam, Joe, Baba and Higgs the Genetically Modified Wolf

… to the moon and back …

and for my Dad

… who gave me my love of wild things.

Contents

Introduction

4913 Penn Avenue, Pittsburgh, Pennsylvania

Four doors down from a gluten-free bakery, two doors up from a Vietnamese takeaway, is one of the world's most unusual museums. Its outside is elegantly understated. There are no ornate pillars or sweeping staircases. Instead, the façade is a muted, minimalist chequerboard fashioned from glass and steel. At a casual glance, it looks more like a trendy boutique than it does a centre for learning, yet Pittsburgh has a hidden gem squirrelled away in this bustling business street. One online review describes the museum as 'wonderfully weird'. Another says, 'there is nothing like it in the world'. And they're right. Welcome to the Center for PostNatural History.

The museum's ambassador is Freckles, a milky-coloured goat that greets the visitors when they arrive. An 'ex'-ungulate, she's stuffed and mounted, portraying a jaunty demeanour that borders on the mischievous. Pert ears are cocked forwards. Olive eyes protrude from a long face and her mouth is turned upwards into the semblance of a smile. Although she looks ordinary from the outside, on the inside she is anything but. Her DNA is part goat, part spider. She doesn't have eight legs and has never spun a web, but when she was alive she did have superpowers. Freckles used to make spider proteins in her milk. She was donated to the museum by the scientists who deliberately modified her DNA. 'She's there as a conversation starter,' says the museum's director, Richard Pell. 'She's the first thing that you see when you walk in the door.'

Spend time at the museum and you'll find it is home to an impressive range of equally idiosyncratic oddities. There are pressed leaves from genetically modified plants, the grimacing skull of a pug and a stuffed salmon that contains the DNA of not one but three different species. There's a fluffy chicken with extra toes, a weird hybrid brine shrimp and the testicles of a cat called Jimmy Cat Carter. If Tim Burton made natural history museums, this is how they would be.

None of these exhibits would be found in a classic natural history museum. 'They have a blind spot for things like this,' says Richard. Conventional natural history museums are full of dinosaurs, glass-eyed tigers and stuffed wild birds. They're designed to showcase the natural world, but Richard's exhibits don't belong there because they're not perceived as natural. They're not 'unnatural' because they are carved from the same biological building blocks as all other living things, but they're not quite 'natural' either because they didn't evolve into their current incarnations on their own. Instead, every single one has been deliberately created by humans. 'They are all living things that have been intentionally altered in ways that are heritable and that change their evolutionary trajectory,' says Richard. So he calls his specimens 'post-natural' to reflect the fact that they were created after the arrival and interference of humans.

This is a book about the Earth's post-natural history. It's about the relationship that exists between humans and other species, and how this has changed over time. From humble beginnings in the cradle of Africa, humans have risen to become a global superpower. Along the way, we invented technology that lets us shape the behaviour and biology of living things. We used this new power to redesign animals, plants and other organisms, and derailed their evolutionary journeys onto a new post-natural path.

Domestic species, such as dogs, cattle, sheep and pigs are all part of this story. Now we live in a world where cloned ponies play in polo tournaments, cattle are being engineered to resist disease and pigs are being modified to grow human transplant organs. The salmon with the extra DNA is a new super speedy-growing variety, while the chicken with the extra digit is one of a baffling array of domestic birds that have been variously produced for their meat, eggs, size, colour and pure novelty value. Extinct creatures are being brought back to life. As Freckles demonstrates, living animals are being modified to produce new materials and medicines. Ranchers are selectively breeding weird-coloured wild animals for paying punters to shoot, people can now pay to have their pet dog or cat cloned, and in the US, it's possible to buy tropical fish with added jellyfish genes that make them dayglow colours. Meanwhile, Jimmy Cat Carter's lonely gonads serve to remind us that while humans have achieved new levels of control over the future of life, we also influence the future of death. 'Humans have acquired many strategies to prevent life from reproducing,' says Richard. Castration is one way, as Jimmy Cat Carter found to his cost, but now researchers are developing methods that could drive entire species to extinction.

All these examples are products of intentional design, but as our global dominance reaches new heights, our influence extends far beyond these lifeforms. As we raze forests, pollute oceans, warm our world and radically alter the biosphere, we are now influencing the evolution of *all* living things, near and far. Although these changes are not calculatedly deliberate, they are still post-natural because humans are at the helm. Our activities are placing the natural world in danger and now many wild species face an uncertain future. Extinction is now an everyday occurrence and even seemingly bulletproof populations of common species are taking a pummelling.

As the pace of environmental change intensifies, evolution is speeding up. Weird hybrids are appearing and new species are beginning to evolve. Polar bears and grizzly bears are giving birth to hybrid 'pizzly' offspring. Narwhals are mating with beluga whales. Mice in New York's Central Park have evolved the ability to digest pizza, while anole lizards in Puerto Rico have evolved stickier toe pads that help them cling to buildings. Hunting is causing elephants to evolve smaller tusks, pollution is driving the emergence of toxin-resistant fish and climate change is causing birds to evolve new plumage.

The timing of these events is no coincidence. They are the repercussions of our actions. They may not be premeditated, but they are far-reaching. Now we live on a planet where humans have become the dominant force sculpting evolution. The world hasn't seen evolutionary change on this scale since the demise of the dinosaurs. Life is changing. Humans are responsible.

The stories I share in this book are predominantly from the animal kingdom, although other walks of life are also available. This reflects my own mammalian bias, so apologies to the various groups that are under-represented. It is a book in three slightly overlapping parts. The first part describes the species we have engineered deliberately. It charts the rise of domestication, selective breeding and modern molecular methods to alter the DNA of living things. It describes our complicated relationship with the natural world and our growing sense of mastery over it. The second part describes some of the species whose evolutionary paths we have modified accidentally. It considers what happens when domestic species start to displace wild ones, and the evolutionary repercussions of humanity's global reach. In the final part, we'll explore some of the methods that are being used to restore the world's lost biodiversity and guide evolution onto more secure footings. These stories are a source of inspiration and of hope, because they show us that, when humans take time to care about natural world, great things are possible.

CHAPTER ONE

The Wolf that Rolled Over

Please don't judge me. I am about to tell you something you may find shocking, and I am concerned you may think badly of me. When I've told people before, it has divided opinion. Some have been curious, others downright disgusted. They've told me it's unnatural and asked me how I could do such a thing. I've no idea how you'll react, so I'll just come straight out and say it.

I own a genetically modified wolf.

I really do. My husband and I got him from a breeder that we found on the Internet. We exchanged a couple of emails, transferred a hefty wedge of cash then collected him from a pre-arranged location in southern England. The little animal howled all the way home.

Five years later, we now trust him so much that he lives in our house, sleeps on our bed and plays with our kids. If we were to set him free, I'm almost certain he wouldn't survive. He's never hunted fresh meat or brought down a caribou. He'd probably hang around by our back door, sulk and wait to be let back in again.

Higgs, as we call him, is a weird-looking wolf. His DNA has been altered so he is less than half the size of his free-roaming ancestors. His skull is smaller, his snout less pointy and his ears flop down rather than standing erect. The classic sleek pelt has been replaced with what can only be described as an embarrassment of soft, messy curls. He is black all over, except for his nose, belly, tail and feet, which are white … or brown when he's been digging in the garden. His tail wags rhythmically when he hears the word 'cheese'. Behaviourally, all trace of wolf cunning has been obliterated. The result is an animal so far removed from its original wild form that he barks at bin bags and often refuses to go out in the rain.

Before you pass judgement on this apparent lupine freak, let me tell you I am not the only one to own a genetically modified wolf. Millions of people, all over the world, keep similar animals but know them by a different name. They call them dogs. For dogs *are* genetically modified wolves.

When people think about genetic modification (GM), they tend to think about animals and plants whose DNA has been sculpted using the modern tools of genetics, but domesticated species have been genetically manipulated too. From the diminutive dachshund to the massive Saint Bernard, all dogs are descended from the European grey wolf. At some point in the past, humans and wolves crossed paths, and then somehow, somewhere, the wolf began to change. Its appearance altered. The wolf began to shrink. Its coat changed colour and its face changed shape. Physiological differences emerged, like the ability to digest starch and give birth more often. Its behaviour changed. The fearsome apex predator morphed from an animal that actively shuns human company into one, like Higgs, that demands it. All of these differences are underpinned by changes to the wolf's genetic code. Now, although wolves and dogs still share around 99.5 per cent of their DNA, the tiny fraction that is different is enough to imbue them with their vastly different features.

Today, dogs have become such a normal part of our lives that it's easy to take them for granted, but their emergence marks a defining moment in the natural history of our world. Dogs were the first domesticated animals. It was the first time humans took a species and then fashioned it to become something more preferable. It was the first time we wrestled control of evolution and began to steer the biology of living things in a different, post-natural direction. The emergence of dogs paved the way for other domesticated species to follow, triggering a chain of cause and effect that would change our world for ever.

According to the most recent estimates, modern humans evolved in Africa sometime between 350,000 and 260,000 years ago, and for the vast majority of the time that followed, we simply lived off the land. We existed as hunter-gatherers, and were entirely dependent on wild animals and plants for our survival. Domestication changed all that. Around 10,000 years ago, after we had domesticated dogs, we began to strike up alliances with other wild organisms. The repeated harvesting and sowing of wild cereals led to the creation of domestic crop strains that were more bountiful and easier to grow. We domesticated other animals, like sheep, cows and goats, and as we began to corral and keep them, and tend to our crops, we found ourselves increasingly tied to the land. The nomadic hunter-gatherer way of life gave way to a more settled existence, leading to the formation of villages. Because they could be bred, domestic animals provided a renewable source of meat and milk for food, and wool and leather for clothing. Food became more plentiful and the population began to grow. In time, because they could be owned and were easily transportable, domestic animals and plants went on to become a source of capital and wealth, so domestication fuelled the rise of trade. It drove the development of new technology, like ploughs, which further accelerated the rise of agriculture and in time led to the development of urban communities. When we think about key innovations, it's all too easy to dwell on recent inventions like the Internet and antibiotics, but it's no understatement to say that domestication helped to fuel the rise of civilisation, and changed the course of human history.

Looking around me now, I see a world full of domesticated species. My genetically modified hound, Higgs, slumbers peacefully at my feet. In the garden outside, our five chickens peck at corn, while our two rabbits nibble on a carrot. There are ponies in the field next door, and sitting on the fence post, a wayward tabby cat eyes me with disdain. Sitting at my desk sipping milky tea,* it's hard to imagine a time when the world was not full of domesticated animals, plants and the products derived from them. Yet, for the vast majority of time that there has been life on Earth, there have been no domesticated animals or plants. So when and where did this momentous change take place?

---

* Dogs, chickens, rabbits, ponies, cats, corn, carrots and tea are all species that have been domesticated.

What Was the Time Mr Wolf?

Until quite recently, scientists thought dogs were domesticated around 15,000 years ago, towards the end of the last Ice Age. It was a time when the ice sheets were retreating, when the landscape was newly green, and when humans and other animals began to colonise the northerly regions of Europe and Asia. There are plenty of dog fossils from this time, found in archaeological sites across Europe, Asia and North America, and the scientists who have studied them all agree: these remains belong to dogs, not wolves. The proportions of their skulls and the shapes of their teeth are all quite different. But then came a fossil that left people scratching their heads.

The skull was discovered in the Goyet Cave in Belgium. It's a remarkable archaeological site jam-packed with the bones of ancient humans, Ice Age animals and other captivating relics. 'The skull is quite small,' says Mietje Germonpré from the Royal Belgian Institute of Natural Sciences in Brussels, who studied the fossil. 'It is about the same size as a modern German Shepherd skull.' Wolves have long, slender snouts, but this animal had a shorter, wider muzzle and a broader braincase. It also had large, primitive-looking teeth. Collectively the features suggested that this animal was more dog than wolf. 'So we decided it was a primitive dog,' she says.

Then came the bombshell. Radiocarbon dating revealed that the skull was actually much older than previously thought. The creature was 36,000 years old, potentially pushing back the origins of domestication by a staggering 21,000 years. 'We were very surprised when we found out,' says Mietje. The skull divided opinion.

Some people agreed with Mietje. Others did not. 'They said it's too old and they don't consider it to be a dog,' she says. Critics pointed out that wolf skulls from this time vary enormously in size and shape and suggested that the Goyet skull belonged to an odd-looking wolf rather than an early dog. Then a different group of researchers made a computer-generated 3D reconstruction of the skull and concluded that certain features, like the way the snout protruded from the skull, were also wolf-like. It could have been an end to the debate, but then other fossils cropped up. Mietje has studied dog-like skulls from the Czech Republic and Russia that are over 25,000 years old, while a separate research group has described the 33,000-year-old skull of a presumed dog found in Siberia's Altai mountains. What to think?

It's bound to be tricky. If these animals really are early dogs, then they're 'only just' dogs so they're bound to have dog- and wolf-like features. So it's here that scientists are turning to another form of historical evidence to help resolve the conundrum: ancient DNA.

Although DNA breaks down after death, sometimes the molecule can be preserved inside fossils, and extracted and studied. This gives scientists another way of studying the transition from wolf to dog. In the early days, genetic analyses painted a confusing picture. One study, for example, compared the full genetic sequences, or genomes, of modern dogs and wolves, to determine that dogs were domesticated between 11,000 and 16,000 years ago. Another study of ancient canids, which focused on a subtype of DNA hidden in the cells' energy-generating mitochondria, suggests a date between 19,000 and 32,000 years ago. The results present a massive discrepancy. On the one hand, they suggest dogs were domesticated around the end of the last Ice Age at a time when agriculture was emerging. On the other, it seems they were established on the other side of the Last Glacial Maximum, the time when the ice sheets were at their greatest reach.

The debate moved on in 2015 after Swedish researchers discovered a fragment of rib protruding from a Siberian riverbank. They originally thought the bone belonged to a reindeer, but DNA analysis later confirmed that it came from a wolf. Radiocarbon dating suggested that the animal died around 35,000 years ago, long before dogs were thought to be domesticated, but then further genetic tests muddied the waters. The ancient wolf seemed to be equally related to both modern domestic dogs and modern wolves, but how could this be if dogs had yet to evolve? The team concluded that the ancient wolf must have lived just after the split between the ancestors of today's dogs and the ancestors of modern wolves. This means an earlier date of domestication, around 35,000 years ago, looks increasingly likely. Then in 2017, a different group of researchers arrived at a similar conclusion, this time using Neolithic dog fossils.

As more studies are added, an early date for the metamorphosis of wolves into dogs looks increasingly likely. Genetic analyses and fossil evidence hint at a deep connection between humans and dogs that stretches back much further than was initially assumed. It predates the rise of agriculture and settled societies, and now researchers find themselves arguing over when and where the transition occurred.

Today, the grey wolf is the only member of the canid family to have paws in both the Old and New Worlds. Its current range encompasses much of Europe, Asia and North America, but in the past, its territory was even greater. This makes it hard to know where to start. We know that dogs cannot have been domesticated in North America, because humans didn't arrive there until well after the Last Glacial Maximum when domestication was already well under way elsewhere, but that still leaves much of the globe to consider. Fossil finds point to Europe and further east to Siberia where the earliest, most primitive dog skulls have been found, but ancient DNA studies throw up alternative scenarios. Some suggest dogs became man's best friend in East Asia, while others hint at origins in Central Asia or the Middle East. Meanwhile, a recent study that compared genetic material from modern and ancient specimens revealed an old, deep split between East Asian and Western Eurasian dogs. The most obvious explanation, according to the study's author, Greger Larson from the University of Oxford, is that domestication occurred in at least two different places. The story of dogs may have no single origin. Dogs could have been domesticated multiple times in multiple places.

What intrigues me most, however, is *how* this relationship began. We can be pretty certain that our ancestors didn't just wake up one day and declare they wanted something that would fetch a stick, yet the process of domestication had to begin somewhere.

Hounds of Love

The winter had been cold and long, but now the sun climbed higher in the sky and leaves were beginning to unfurl. The youngster sat on his haunches, staring into the embers of a fire that was dying down. He felt resentful. Not quite a boy, but not yet a man, he had been left behind in camp while the adults went out to search for food. Now he found himself alone, contemplating mischief.

A few days earlier he had followed his father out of the camp and into the woods. His father showed him a place where a large tree had fallen, wrenching deep-seated roots out of the rust-coloured earth, and the resultant hole that had been exposed in the hillside. Sharp-clawed paw prints framed the entrance to a dark tunnel: the unmistakable signature of a she-wolf's den. 'Be careful,' his father had warned him. 'These animals are dangerous.'

Back in camp, the boy knew he had hours before the hunting party would return, so he picked up a spear and slipped out of camp. He returned to the den to find fresh scats on the ground outside. The mother wolf had been there but when she heard his clumsy footsteps, she had beaten a swift retreat. Now she hovered in the background, watching as the human dropped to his knees and plunged his arm deep into the lair. When he stood back up, he was holding a small, wriggling wolf cub. It squirmed and whined, making the boy tighten his grip. Then he swaddled his find in a reindeer hide and carried it back to camp.

* * *

Although humans and wolves have shared the same landscape for many tens of thousands of years, they interacted little. Both would have been wary of the other and kept their distance, but then something must have changed. Academics argue over the nature of this initial interaction, but one scenario is that humans actively decided to invite the wolf into their world. Someone, like our Palaeolithic boy, went out and collected a cub. Skilled hunter-gatherers with an in-depth knowledge of their local environment, they would have known where the wolf dens were. It wouldn't have been difficult to scoop one up and bring it back to camp. Then, having done it once, it would be all too easy to repeat the process. The cubs that were kept would inevitably have been the ones that were easiest to catch, so over time, as the genes for their more relaxed nature were passed through the generations, domestication got under way.

Back in camp, the animals would have been kept for pragmatic purposes. As cubs, they could have entertained the children. As adults, they could have acted as sentries, and if they ever got too boisterous or aggressive to look after, they could have been set free or killed for their meat and fur.

We certainly know that Palaeolithic people wore specialised cold-weather clothing, including a variety of fitted garments made from well-tanned pliable hides. A 24,000-year-old ivory figurine from southern Siberia, for example, depicts what seems to be an individual wearing a carefully tailored all-in-one fur suit. Evidence for a wolf-fur onesie? It's a possibility. Similarly ancient wolf bones have been found with distinctive cut marks, indicating that the animals were probably skinned for their fur. They may also have held symbolic significance. One skull, studied by Mietje, is interesting because it has a bit of mammoth bone wedged between its front teeth. The fragment must have been inserted into the animal's mouth after it died, suggesting human intervention, while other skulls sport conker-sized holes where their brains were removed. There were easier meals to be had than brain, so Mietje thinks these unusual relics are evidence that dogs held special significance. 'I'm in favour of the active involvement of the Palaeolithic people,' she says. 'I think they actively started to collect these animals and then kept them, not just for their fur, but for rituals too.'

It is, perhaps, easiest to imagine that humans chose the wolf, and that the wolf had no option but to go along with our plans. As a species, we like to think we are superior and separate from the animal kingdom, when really we're just animals too. Today, if we want a dog, we can just go out and get one, but it would be naive to presume that our ancestors followed the same thought process.

So an alternative theory proposes, not that humans chose wolves, but that wolves chose humans. Leftovers discarded by humans lured the wolves out of the shadows. The animals that were least afraid of us were the ones most likely to enter our campsites. As a result, they were better fed, healthier, and more likely to reproduce than warier pack members. The genes underpinning their more relaxed nature were passed between generations, and over time, the animals became progressively tamer. In this 'self-domestication' scenario, humans were stooges. We didn't invite wolves in, but by being messy, we created an ecological niche they were only too happy to fill.

It's a possibility. Modern wolves are adaptable animals. In Canada, there are two types of wolf: 'nomadic' wolves that follow the caribou around and 'sedentary' wolves that tend to stay in one place. From time to time, their paths cross, but they don't really get on. They're like the Starks and the Lannisters, and will fight each other to protect what they consider to be *their* caribou. So maybe, 35,000 years ago there were one or more groups of migratory wolves that considered us to be their property. Instead of tracking caribou or reindeer, they followed us around, not because they wanted to eat us but because they benefited from the shared association.

So which was it? Did humans choose wolves or did wolves choose humans? We'll probably never know but the upshot was the same. After contact was made, humans and wolves began to interact and over time, the relationship strengthened. Primitive dogs probably accompanied humans on their hunts and so tipped the odds in favour of a kill – a reciprocal arrangement that benefited both parties. At some point, when we started to physically keep them with us, we would have started engineering which animals got to reproduce. In the early days, it would have been the calmer animals that would have tolerated living in captivity, but in later times we would have selected for other characteristics, like being a good sentry or scaring the neighbours. It was the beginning of a long and beautiful friendship, and a defining moment in the story of evolution. Of course, dogs were only the beginning …

Pigeon Parade

If Charles Darwin is to be remembered for only one book, it has to be the now classic *On the Origin of Species by Means of Natural Selection*. Published in 1859, this highly readable doorstop outlines the great man's theory of evolution by natural selection. According to the theory, individual members of the same species are similar but slightly different. These variations make the individual more or less suited to the surrounding environment. The best adapted or 'fittest' individuals are more likely to reproduce and pass their winning characteristics on to future generations, while those less suited are more likely to die out without any such legacy. Selection – where certain characteristics are favoured over others – is driven by natural forces, like the availability of food, enabling species to change and evolve over time. Darwin cited evidence gathered from his time on HMS *Beagle*, including, most famously, his work on the Galápagos finches, but does *Origins* start with the story of Darwin's finches? No, it does not. Instead, Darwin focused on a different passion: his beloved pigeons.

At the time, pigeon-fancying was all the rage in Victorian England, and Darwin kept a loft-full in his garden at Down House. He wrote that spending time with them was 'the greatest treat which can be offered to a human being', and loved them so much that when his daughter's cat did what cats do, and killed a couple, he had the miscreant moggy shot. So much for survival of the fittest; if only the cat had had time to evolve a Kevlar coat! It's said that his daughter, Henrietta, never forgave him, but Darwin was blinkered. He bought as many different varieties as he could. There was the 'Pouter' with its balloon-like crop, the 'Jacobin' with its luxurious 'feather boa' collar, and the short-faced tumbler, which flies to great heights then somersaults out of the sky, to name but a few. Darwin realised that although the birds were incredibly different, they must all have descended from the same common ancestor, the rock pigeon. Instead of being sculpted by natural forces, Darwin thought that the birds' idiosyncrasies must have been shaped by artificial means, namely the whim of the pigeon-fanciers who bred them.

Then in 1868, he published a lesser-known tome, *The Variation of Animals and Plants under Domestication*, which explored domestication more deeply. One of the things he noticed is that domesticated animals share a suite of characteristics not seen in their wild forebears. They tend to be smaller, with patchy markings, curly tails and floppy ears. Think about it for a moment … can you name a wild animal that is piebald? I know we have zebras and pandas and skunks and orcas and magpies, but their markings are more ordered than the irregular patches of a Holstein cow or a collie dog. How about ears? Can you name a wild species that has droopy, spaniel-like ears? Elephant ears don't count. They're flappy rather than floppy. I'm struggling. Darwin concluded, 'Not a single domestic animal can be named which has not in some country drooping ears.'* Sometimes, the proportions of body parts are altered. Compared with a wolf, for example, the English bulldog has a short skull and muzzle with a pronounced underbite – the bottom jaw sticks out further than the top one, giving the breed its iconic

* There are, of course, exceptions. I've yet to meet a goldfish with spaniel ears, but the colourful creatures can differ greatly in size from their wild ancestor, the carp, and do sometimes have piebald markings, both tell-tale signs of domestication.

grimace. Pigs have extra bones in their spine and longer bodies than their wild boar ancestors. The reproductive habits of domesticated animals are also often changed, with many species able to breed all year round.

Darwin recognised this hodgepodge of shared features – the so-called Domestication Syndrome – but couldn't fathom how it came about. Indeed it's been called one of the oldest conundrums in genetics. Darwin wasn't sure if these features had been deliberately selected for – if our distant ancestors deliberately chose floppy-eared animals and bred them together, for example – or whether they came about by chance. And although he could experiment with species that were already domesticated, like his beloved pigeons, he thought that domestication happened so slowly it would be impossible to study directly. If only there was a way to study domestication in real time.

The Silver Fox

Autumn. A gentle breeze caressed the golden leaves as blackbirds plundered the last of the berries from tangled brambles. The fox was trying to ignore me. An inquisitive blur of perpetual motion, the delicate animal scoured the hedgerow, then headed into the open field. I was just a dead weight on the other end of an extendable lead. The fox kept a respectful distance: close enough so he didn't overstretch the cord that connected us, but far enough away that I wouldn't impede his explorations. He wheeled around me, tracing a series of never-ending circles.

'You'll have to spin round with him so you don't get tangled up,' Emma advised. 'I'd rather you didn't switch hands, because there's always the chance you might drop the lead.'

'What would happen if I did?' I asked.

'Oh, he probably wouldn't go anywhere,' she answered, 'not while I've got the treats.' She held up a pouch full of chicken pieces.

It's not every day you are given the opportunity to take a silver fox for a walk, so when the owners of Heythrop Zoo offered, I jumped at the chance. Silver foxes are a melanistic version of the red fox. This means that although they share many of the red fox's features, including the pinched muzzle and long, bushy tail, their colouring is different. Silver foxes are, as the name suggests, varying shades of grey and not red at all.

My fox,* Glacier, is particularly beautiful. Olive eyes peer quizzically from a face framed in silver-flecked fur. Dainty black legs support a compact body draped in a dense, luxurious coat that is dark at the base and silver at the tip. His thick, bushy tail is silver too, but it looks like the end has been dipped in a pot of white emulsion paint.

'You can touch him if you like,' says Emma.

To attempt this with a silver fox straight out of the wild would be nothing short of disaster. The animals have speedy reactions and very sharp teeth, but Glacier is different. Heythrop Zoo, one of the largest private animal collections in Europe, provides trained animals to the media, for use in films, commercials, music videos and the like. Glacier is one of them. He was donated to the

---

* I say 'my fox', although sadly Heythrop would not let me take him home.

zoo by private owners, who reared him from a cub, then took him into schools for educational purposes. Things went well until, at around a year of age, Glacier's hormones kicked in and the adolescent fox cub became harder to manage. Kids became nervous, bookings dropped off, and the couple found Glacier increasingly difficult to manage. So Heythrop took him in.

He now lives in the Cotswolds in a large, outdoor enclosure that he shares with a handful of other regular foxes. He's built up a trusting relationship with Emma Hills, one of the zoo's specialist animal trainers, who has spent hundreds of hours working with him. She applies the principles of American psychologist B. F. Skinner, where positive behaviours are rewarded, and negative behaviours are ignored. So Glacier has learned that putting his head in a collar or sitting on command earns him a piece of chicken. Everybody's happy.

'He always has the choice whether or not to interact with us,' Emma explains. 'When I present him with the collar and lead, he decides whether he wants to come for a walk or not. If he backs away, we leave it for another day. If he puts his head into the collar, he gets a reward and then we go and explore. If you raise your finger in the air, he knows to sit and if you say the word "touching", he gets to choose whether or not you touch him. If he stays sitting, you can stroke him. If he backs away then you leave him alone.'

Watching the busy creature whirling on the end of the lead, I found it hard to imagine Glacier paying any attention to me at all, but I gave it a go.

'Glacier,' I called in a sing-song voice. The little fox stopped immediately and eyed me inquisitively. Then I raised my index finger and he sat. 'Touching,' I said, then when Glacier showed no signs of moving away, I bent down and ran my hand down his silver-grey back. It is the smoothest, glossiest, thickest fur coat that I have ever felt. I offered him a piece of chicken and our moment was concluded. He returned to his explorations, and we continued on our way.

Glacier is clearly not a domesticated animal, yet somehow he's not quite wild either. He lives in captivity and goes for walks on a lead, yet I certainly wouldn't trust him not to snaffle the chickens from my garden. He's trained but he's not exactly what you'd call tame. He interacts with humans but you get the feeling that he'd much rather be with foxes. Glacier is an enigma wrapped up in a conundrum bundled up in the world's most beautiful fur coat. Yet the silver fox has taught us more about the transition from wolf to dog, and indeed about the process of domestication, than any other animal.

The Fox that Rolled Over

It was a dangerous time to be a geneticist. Although the Soviet dictator Joseph Stalin was dead, his legacy cast a long shadow. In the Great Purge of 1936 to 1938, Stalin had tried to stamp out the corruption he thought pervaded Soviet society. Under his orders, 1.3 million 'saboteurs' and 'counter-revolutionaries' were arrested, of whom half were killed. Paranoia and mass murder ruled. Then, when the Second World War ended and the Cold War began, relationships between the Soviet Union and the 'capitalist' West disintegrated. Stalin denounced Western ideals. Genetics was banned and research laboratories were shut down. Practising scientists and open supporters of Darwin were declared enemies of the state. If you were lucky, this meant losing your job. If you were unlucky, it meant losing your life.

It was against this backdrop of repression, violence and fear that a Russian scientist decided to set up one of the greatest genetics experiments of all time. That scientist was Dmitri Belyaev.* Like Darwin, Belyaev was fascinated by evolution and the process that carved wild animals into their myriad domestic forms. Unlike Darwin, however, he didn't think that domestication had to be so painfully slow. Breeding experiments with mink had led him to believe that features of the domestication syndrome could occur relatively quickly. So he decided to try to domesticate an animal from scratch so he could study how the process occurred.

The silver fox was an obvious choice. A member of the canid family – the group of carnivorous mammals that includes dogs and wolves – it's an intelligent, social animal that lives in small family groups. Unlike the dog, however, it had never been domesticated. When he began his experiment in 1959, the fur trade was big business. Silver foxes were farmed on an industrial scale and widely exported. This provided him with a ready supply of test subjects, and a smokescreen for his controversial genetic research. He told the Russian authorities that his breeding experiments were to improve the quality of the animals' pelts.

---

* Belyaev was a remarkable man, not least because when he started his experiments, his geneticist brother, Nikolai, had already been executed by the secret police.

He set up his experiment in the wilds of Siberia in a place called Novosibirsk. It was about as far away from prying eyes as was physically possible. He sourced his foxes from fur farms where the animals had been bred for around 50 years, but the animals were by no means domesticated. Most of the initial batch of 130 foxes would bare their teeth and lunge at him, but a small subset – around 10 per cent – were marginally less aggressive, so he chose these as the founders for his breeding experiment. The foxes were never fussed or petted. The idea was not to tame or train them, as Glacier has been trained, but to select the animals that were least fearful and allow them to breed. Then, when those animals grew up, the process was repeated. The least aggressive foxes became parents, while the more aggressive foxes became fur coats. Like any good scientist, Belyaev also set up a control group; a second skulk* of foxes that were allowed to breed at random. This gave Belyaev a yardstick against which to measure change. His idea was that when our ancestors started domesticating dogs and other animals, they too must have begun by selecting the tamest individuals. And if tameness had a genetic component, which Belyaev felt sure was the case, then over time he expected tameness to become more pronounced.

Change came thick and fast. By the fourth generation, foxes from the experimental group had started to wag their tails. By the sixth generation, they were licking the scientists' faces. The proportion of amicable foxes in each

* 'Skulk' is the collective noun for foxes. Other favourites include a bloat of hippos and a smack of jellyfish.

litter grew, until by generation 45, virtually all of the foxes were behaving like friendly dogs. Physical changes began to happen too. After just 10 generations, some of the foxes were starting to develop floppy ears, curly tails and piebald coats. Their classic argentine fur became peppered with white patches. They began to breed more often and their skeletons began to change. Their legs were shorter, their snouts became smaller and their skulls became broader, giving them a less vulpine, more dog-like appearance.

Today, a visit to the fox farm is a joyful encounter. Now *all* of the foxes in the experimental group are super-friendly. Geneticist Anna Kukekova from the University of Illinois visits the Novosibirsk farm at least once a year as part of her research. 'It's really fascinating to see these animals,' she says. 'They live in rows of cages in a big shed, and when you walk in, they're just so friendly. You read that they are tame, but you do not realise how happy they are to see people. These are grown-up animals but they act like little puppies.' There's even evidence to suggest that the foxes have started to *think* more like dogs too. Dogs are well known for their ability to interpret human gestures. Hide a treat under one of two upturned cups and a dog can infer where it is by watching as a person points or gazes at the correct location. Remarkably, the tame fox kits can do this too. They are every bit as good at deciphering human gestures as regular dog puppies and run rings around kits from the control group.

To make sure his results weren't just some quirk of fox biology, Belyaev tried using the same approach to tame other species. Starting in the seventies, he repeated the experiment with Siberian grey rats, mink and river otters, all with broadly parallel results. As the proportion of tame animals grew across generations, changes in colour, anatomy and physiology followed.

The experiments all point towards the same broad conclusion. Domestication doesn't have to take the achingly long time periods that Darwin envisaged. It can be replicated quite quickly in laboratory settings, by doing just one thing: selecting for tameness. Consistently breeding the friendliest animals, year in, year out, is all it takes to produce the seemingly unrelated mishmash of features seen in the domestication syndrome. Our distant ancestors didn't necessarily select dogs on the basis of their floppy ears or ability to read our body language; instead, these skills could have occurred as a by-product of selecting for their friendly nature. We didn't deliberately engineer the domestication syndrome. If you like dogs with floppy ears, which I do, you can consider it a happy accident.

Panimals

In 2004, scientists put forward a theory to explain *why* repeatedly breeding friendly animals together leads to the features of the domestication syndrome. Tecumseh Fitch from the University of Vienna and his colleagues believe the answer lies in a group of cells called the neural crest. Early in life, when vertebrate embryos are developing, cells from the neural crest migrate away to different parts of the body where they go on to form various different cells and tissue types. They help to form ears, teeth and pigment-producing cells. They also contribute to the adrenal gland, which helps prepare the body for 'fight or flight'. And in dogs, they travel all the way down the length of the body to the tail, where they help give the appendage its characteristic size and structure.

The idea is that, as tameness is selected for over generations, something happens to the neural crest. Neural crest cells start migrating but many of them never make it to their final destination. The ears don't get to form properly, so they tend to flop rather than stand proud. The snout doesn't fully extend, resulting in a shorter dog-like muzzle. There are changes to the tail, making it curly rather than straight, and pigment-producing cells don't mature properly, creating patchy coloured splodges of fur. Critically, the adrenal gland never gets to fully mature, so the fear response is dampened. The end result is an animal that never quite matures properly, physically or mentally.

This seems to fit. It's as if domestication derails normal development and suspends animals in a state of frozen youthfulness. Belyaev's tame adult foxes behave like wild fox kits. Dogs are essentially wolf puppies that never really got the chance to grow up. Domesticated animals are the Peter Pan of the animal kingdom or, as I prefer to call them, 'Panimals'.

When Darwin put forward his theory of evolution, he realised that certain characteristics were inherited, but he didn't know how. Today, we know it's DNA that is passed between generations. Random changes to the molecule – mutations – provide the source material for natural selection to act on. Advantageous mutations are favoured, while unfavourable ones are weeded out. When our ancestors began interbreeding the least wary animals, and kick-started the process of domestication, they were unwittingly selecting for the genetic mutations that underpin a docile temperament. Somewhere along the way, it's likely that one or more neural crest-related genes acquired a mutation and were also inadvertently selected, leading to the features of the domestication syndrome.

Anna Kukekova has been searching for the relevant mutations by studying DNA from Belyaev's foxes and so far has highlighted more than 100 relevant genes. This is to be expected. Domestication, with its varied physical, behavioural and biological components, is likely to have a complex genetic basis. As predicted, some of the genes relate to the neural crest, but others are involved in brain function and potentially behaviour. This is important. 'When you visit the animals, it's not the physical changes, but the behavioural changes that stand out,' says Anna. One of the genes – called *SorCS1* – is known to influence the way that neurons communicate with one another inside the brain. Here then, at last, is a suite of genetic changes that could help to explain the behavioural transition from wild to tame.

The Novosibirsk fox-breeding experiment continues to this day. When Belyaev died in 1985, his long-time assistant Lyudmila Trut took over. Times became hard after the fall of the Soviet Union, so many of the experiments were scaled down and the Institute for Cytology and Genetics began selling the tame foxes as pets. According to Trut, a number of the animals now live happily with families in Russia, Western Europe and North America. From time to time, the owners write to Trut to tell her how the foxes are doing. She gets letters from their owners, who tell her how sociable and affectionate their pets are, but these are not easy animals to keep.

Owing to vaccination regulations, the foxes can only be exported to the US or Europe when they are over six months old. This means that the early, formative parts of their life are unavoidably spent behind bars in the Novosibirsk sheds. As a result, the kits become adapted to their cages and enclosures, then struggle to adapt to life 'on the outside'. Like so many of my former boyfriends, they also have a strong odour, are annoyingly overactive at night, and can be difficult to housetrain. Often, they don't take well to apartment-style living and have to be kept caged, or under close supervision on the end of a lead. 'They were selected to be friendly to humans but they were not selected to fit the human lifestyle,' says Anna. Although you can take the fox out of the wild, it seems difficult to fully take the wild out of the fox. Although these animals have changed immensely over the course of the breeding experiment, whether or not they are fully domesticated is an issue that remains contentious.

Are We Nearly There Yet?

By selecting for tameness over generations, Belyaev pressed the fast-forward button on domestication, but it would be unrealistic to expect that our ancestors did the same thing. In laboratory conditions, domestication *can* happen relatively quickly, but Darwin was right when he surmised that the process usually takes much longer. Just as we can't see one species evolve into another, so too we are unable to witness the moment when wild becomes domestic. There was no instant in time when the wolf suddenly stopped being a wolf and started being a dog; instead the changes would have occurred so slowly it's unlikely our ancestors realised what was happening.

When I first started researching domestication, I thought that defining it would be as simple as pointing at some animals and saying 'chicken, cow, sheep', yet it's now obvious to me that the boundary between wild and tame is blurry and indistinct. Belyaev's foxes illustrate this beautifully, as does the so-called 'domestic' cat. It's thought the famously independent creature began to domesticate itself after agriculture took off in the Middle East around 9,500 years ago. When vermin started ransacking grain stores, wild cats reinvented themselves as pest control. Compared to dogs, however, cats have changed little on their journey from the wild. They still look and act similarly to their wild ancestors. They hunt wild food and they have wild sex. In Scotland, domestic cats are interbreeding with the endangered Scottish wildcat, an act that is diluting the wildcat's gene pool and nudging it towards extinction. The domestic cat's infamous sangfroid nature arguably suggests another species whose domestication is a work in progress.

In Mongolia, native domestic horses roam free and are only caught when needed. Descended from the war steeds of the legendary Genghis Khan, the horses graze the vast grasslands of the Eurasian Steppe, followed at a distance by the nomadic Mongols who own them. Sometimes, the domestic horses interbreed with wild horses, and sometimes feral horses are caught and added to the herd. The domestic Mongol horses are among the most genetically diverse of all horses, meaning that across all individuals there is a healthy smattering of genetic variation, and that humans have interfered little in the breeding process. Their journey through the generations has been dominated by natural rather than artificial selection, yet we arrive at a species that is still considered domestic. A nomadic Mongol herder wouldn't consider his horses to be wild. They are his property. He milks them. He rides them. He races them. Sometimes, he even boils them up in a stew. But a junior member of the British Riding Club, who has grown up plaiting the mane of her infinitely more domesticated steed, would run screaming if asked to mount the same saddle-free spirit.

Textbook definitions suggest that, alongside certain other features, domesticated animals should have an inherited tameness. By this yardstick, Glacier, the silver fox I met at Heythrop Zoo, is not domesticated because although he is relatively manageable, he did not acquire his disposition from his parents. Although his provenance is uncertain, it's thought Glacier is descended from fur-farm stock rather than Belyaev's epic experiment. So his amenable manner is not down to his DNA; it's down to his trainer, Emma. My dog, on the other hand, was born soppy. But what of the millions of stray dogs that live wild? They may well be genetically distinct from wolves, have curly tails and piebald markings, but they're certainly not pets. They don't live with us or depend on us directly for food, but they're not exactly wild either.

With their unique DNA, unusual looks and inherited bonhomie, Belyaev's silver foxes must, at the very least, be well on the way to becoming domesticated. The

Russian's experiment has shown us that domestication can occur relatively quickly and that tameness goes hand in hand with other, more visible changes, but it also teaches us that domestication itself is frustratingly hard to pin down. This should not come as a surprise. There is no 'moment' when a wild animal becomes domesticated; rather, the animals exist on a never-ending continuum. Domestication is a process rather than an event, or, if you prefer hackneyed self-help rhetoric, it's a journey, not a destination. Every generation is subtly different from the one that precedes it, and it's only with the benefit of palaeontological and genetic hindsight that we see grand changes in operation. It's easy to imagine that the dogs and cows and chickens we see around us today are the zenith of domestication, when in reality, they are still in transition. Living things don't stop responding to their environment and evolution never stands still. Domestication is just a human-influenced form of evolution.

In addition, our conceptions of the past are tainted by cultural bias. We imagine that early dogs looked 'dog-like' but our conception of what 'dog-like' means is tainted by the cultural space that we occupy. Let me give you an example. The British Museum hosts an etching by William Hogarth entitled *Gulielmus Hogarth*. This exquisite self-portrait, created in 1749, shows the English painter – a dead ringer for John Malkovich – posing with his beloved pet pug, Trump. But this is not the bug-eyed, scrotum-faced lapdog of your twenty-first-century preconception. It's an altogether different dog, muscly and wizened with a wrinkle-free face that protrudes in three rather than two dimensions, and eyes that sit comfortably in its skull. Back then, the breed was celebrated, not as a family pet, but for its feisty 'pugnacious' spirit, so it held symbolic importance for Hogarth, who used the dog as an emblem in his career. To Hogarth, Trump must have seemed the absolute pinnacle of domestication, the very epitome of 'pugness', yet to us this dog looks somehow primitive and unworthy of the pug moniker. So is it a pug or not? I would argue that it is a pug, but add we must remember that over time, the animals we gave this arbitrary label to have sported a whole range of different features. 'We constantly map onto the past what we think about the present,' says Greger Larson from Oxford University, who studies domestication, 'but we're always wrong because things are always changing through time.'

All this makes it difficult for those who seek to define domestication or pinpoint its inception. For most of its history, domestication was actually a very loose arrangement. 'What we're talking about here is a long-term process by which people, plants and animals become acclimated to one another,' says Greger. Animals got used to people, and people got used to animals. We just hung out. Then, as we started to influence their environment by feeding them, keeping them and moving them around, the animals began to change. Domestication got under way. It wasn't intentional. 'Nobody set out to domesticate a cow,' he says. 'Nobody set out to domesticate a sheep or a pig or a dog or anything else. Instead, domestication was a side effect of the tightening relationship between humans and animals.'

As new forms of animals and plants began to emerge, domestication changed our world. It led to the emergence of agriculture, trade and settled urban living. The process steered animals and plants away from one evolutionary trajectory, towards another. Domestication has had a huge influence on the course of human history, and yet it's intriguing to think that the process was never actually planned or premeditated.

CHAPTER TWO

Strategic Moos and Golden Gnus

In 2018, a giant cow called Knickers made headlines around the world when he was photographed with his herd on a farm in Australia. The black and white behemoth stood horns and shoulders above his slightly cowed companions, making him look like that giant kid at the back of the school photo and them look like miniatures from a children's playset. At around two metres (6ft 7in) tall, Knickers weighed a staggering 1,270kg (2,800lb), which is equivalent to eight fat Elvis Presleys or 11,200 quarter-pounders. The media was amazed by the animal's unusual physique and resultant good fortune. The seven-year-old Holstein was saved from the chop when he grew too big to be put through the local meat-processing facility. Now he has been put out to pasture with his chums on a Lake Preston feedlot in Myalup.

Size is important. Knickers raised eyebrows because he was just so big. Holstein cows are bred as dairy animals, but castrated males, known as steers, are often raised for a few years then sold for beef. At this stage they normally weigh around 700kg (1,500lb), but could feasibly grow much larger if given the chance. By the time he was abattoir-age, Knickers was already looking down on his peers, prompting his curious owner, Geoff Pearson, to hang on to him. Maybe it was his DNA, maybe it was because he was given the chance to keep on growing, but Knickers won his 15 minutes of fame because he was a photographic bovine colossus.

Now cast your mind back 20,000 years ago to a time when bovines of this stature were commonplace. The aurochs is the ancestor of all of today's 1.4 billion domesticated cattle. A formidable beast, it sported a black coat and intimidating forward-facing horns that could grow over a metre (3ft 3in) long. In the sixth book of his *Gallic Wars*, Julius Caesar described the aurochs or *uri* of south-west Europe, saying 'they are a little below the elephant in size, and of the appearance, colour, and shape of a bull. Their strength and speed are extraordinary; they spare neither man nor wild beast which they have espied.'

Towards the end of the last Ice Age, aurochs could be found all the way from the British Isles and central Europe into East Asia and North Africa. For our Neolithic ancestors, armed only with spears, they must have been a source of trepidation and temptation. It would have been dangerous to hunt them, but the rewards – an enormous cache of meat, bones, blood, fat and leather – must have made it a risk worth taking. There's a fabulous archaeological site near the River Tjonger in the Dutch province of Friesland. Fifteen years ago, an amateur archaeologist who was working there stumbled across 49 aurochs bones and a couple of flints. The well-preserved bones came from the spinal column, rib cage and legs of a female aurochs that lived and died around 7,000 years ago. Many of the bones have marks from where the carcass was dismembered and meat was stripped from the bone. The two flint fragments fit together to form an almost complete blade that was probably used for the job, and all of the remnants show signs of having been burned. Nomadic hunters brought down an aurochs, then had a barbecue to celebrate.

Meanwhile, further east, people had already started domesticating our future farm animals. Goats, sheep and pigs were all domesticated in the Middle East between 10,000 and 11,000 years ago, while cattle were domesticated twice: once in the Middle East, around 10,300 years ago, and once in South Asia, around 8,000 years ago. Aurochs from the Fertile Crescent went on to become the European taurine cattle, of which Knickers is a member, while aurochs from the Indus Valley evolved into indicine or humped cattle, also known as zebu.

It was the beginning of a beautiful partnership. To our prehistoric ancestors, these first domesticated cattle must have been the Swiss army knives of the animal kingdom. So many uses! The milk, blood, fat and meat could be consumed, while the hide, horns, hooves and bones could be turned into clothing and tools. The dung could be burned for fuel, and if you could sling a harness around one, the animals could be used to pull a plough or cart. Over the next couple of thousand years we started breeding and castrating them, and taking them with us as we journeyed around the world. Domesticated cattle spread through Asia, Africa, Europe and later the New World. Sometimes taurine cattle bumped into zebu, leading to the emergence of hybrid breeds. The Sanga breed of sub-Saharan Africa appeared in the second millennium BC when local humpless taurines interbred with the first humped zebu.

In its early stages, domestication would have been a loose and informal arrangement, just as it was when our ancestors first started interacting with wolves. Cattle were useful to have around, but there was no strategic, long-term goal. Neolithic people had no vision of a black and white Holstein pumping out milk by the bucketful. How could they? Instead, our ancestors sculpted the form, behaviour and DNA of cattle haplessly. Where we went, they went. Changes happened along the way, but they were by-products of our relationship rather than pre-determined goals.

The fossil record, for example, reveals that domestic cattle actually shrank for much of their evolutionary history. European cattle diminished in size right through the Neolithic period and on into the Middle Ages. Between 5600 and 1500 BC, they diminished by a third. If we'd been directing breeding with forethought, this never would have happened. Instead, it's thought the animals shrank because our ancestors were in a rush to slaughter young adult animals for meat. Neolithic cattle were probably calving at a younger age before they were fully grown, leading to the production of smaller, lighter calves, which then grew up to be smaller, lighter adults.

By the time the Ancient Egyptians were building their pyramids, distinct types of cattle had already emerged. Leggy, long-horned white cattle were used as draft animals; rotund, short-horned cattle were kept for fattening; while slim, fine-boned cattle were used for milk. In a domesticated idyll, cattle no longer needed large horns to defend themselves, so hornless or 'polled' varieties emerged. In recognition of their cultural significance, these polled cattle even had their own hieroglyph.

Strategic Moos

At some point, people started to think very carefully about which animals should be allowed to mate. By the 1700s, geographical variation in cattle herds became apparent. Cows from the Scottish Highlands, for example, looked different from cows from the Norfolk lowlands, but the differences were accidental in the sense that no one had explicitly bred the animals to be different. Up until that point, livestock of both sexes were kept in the same field and allowed to choose their own mates. Then, in the eighteenth century, a British agriculturalist called Robert Bakewell had the brainwave of separating males and females, then deliberately breeding key individuals together.

As obvious as it seems to us now, at the time this was a huge conceptual leap forwards. Breeding was no longer random. Instead it became propelled by the whims and desires of humans. We began, not just to drive evolution along a different route, but to deliberately mould it, enhancing features that we valued. We call it selective breeding. Bakewell noticed, for example, that Longhorn cattle, which were traditionally a beast of burden, were also efficient meat producers. They seemed to eat less while piling on the pounds more easily than other breeds. So he took the 'beefiest' Longhorns and bred them together, then repeated the process with their offspring, and so on. He showed that features such as meaty buttocks for cattle, or lanolin-rich wool for sheep, could be enhanced by the process and thanks to his efforts, the average weight of bulls sold at market more than doubled over 80 years. Pretty soon, other farmers followed suit and selective breeding became a mainstay of agricultural practice.

His ideas did not go unnoticed by Charles Darwin. Bakewell gets a name-check in *Origins*, where Darwin calls the breeds that he enhanced 'a striking example of the power of artificial selection'. Artificial selection, which is really just another name for selective breeding, refers to the fact that artificial rather than natural forces are guiding evolution. These 'artificial' sources are humans.

Over the last few hundred years, selective breeding has intensified, leading to the emergence of specialised breeds, such as the French Limousin and the British Hereford. Sometimes bulls from one country were crossbred with cows from another. Cattle from central Europe were used to boost the milk production of animals in Russia, while hardy Scottish Ayrshires were used to toughen up various Scandinavian breeds. Animals that were naturally polled were interbred to create hornless breeds. In 1824, William McCombie began interbreeding some of the polled, predominantly black cattle of north-east Scotland that were known locally as 'doddies' and 'hummlies'. The result was the Aberdeen Angus, a small, hornless breed of cattle that remains the UK's most popular beef breed. Breed registries were established. They set up and maintained official herd books: bovine databases of *Who's Who* … or *Who's Moo*, if you prefer. Still a vital part of the cattle business today, these records detail exactly how members of the breed are related to one another, so the pedigree of any animal can be traced back as far as the records allow.

In time, this resource helped people to figure out how certain characteristics are inherited, leading to ever more nuanced cycles of selective breeding. When two polled cattle are allowed to breed, for example, it's very rare they produce a horned calf. This is because the polled gene is a textbook example of classic Mendelian genetics. Genes, we now know, come in pairs and are inherited in discrete units, one from each parent. In a series of exceptionally monotonous pea experiments back in the 1860s, Austrian monk Gregor Mendel showed that some genes are dominant, meaning it only takes one copy of the gene to have an effect, while others are recessive, so need two copies to generate an outcome. In the early 1900s, scientists figured out that in cattle, the polled version of the gene is dominant and the horned version is recessive. A calf need only inherit one copy of the polled gene in order to be horn-free.*

The polled version of the gene is a naturally occurring blip or 'mutation' that crops up spontaneously in the animals' genetic makeup from time to time. Darwin was aware of the phenomenon. In his 1868 book, *The Variation of Animals and Plants under Domestication*, he described the appearance of hornless cattle as 'spontaneous variation'. Archaeological evidence reveals that polled animals could be found as far back as Neolithic

---

* If both parents carry two copies of the polled gene, then all of the offspring will inherit two copies and so be polled. If both parents carry one copy of the polled gene each, then 75% of their offspring will be polled.

times, but the gene only became more common when humans decided to deliberately select for the characteristic it determines: being hornless.

It's a similar story with a gene called myostatin that normally puts a damper on muscle growth. It limits the number of muscle fibres present at birth, imposing a natural physical limit to the animal's physique. Sometimes, however, mutations crop up. In 1807, English agriculturalist George Culley described cattle that were unusually well muscled. Put in today's parlance, they were ripped. Although he could not have realised it at the time, we know now that this unusual build is caused by a spelling mistake in the animals' genetic code. Cattle with the mutation have twice as many muscle fibres as their normal counterparts, so the phenomenon has become known as 'double-muscling'. Double-muscled breeds, such as the Piedmontese and the Parthenais, appeared in the late 1800s and as time went on, breeders sought to capitalise on their muscly bulk by crossbreeding these Adonis-like animals with other breeds.*

The Belgian Blue is one such beast. It emerged in the 1950s, post-Second World War, when artificial selection stepped up yet another gear. With the post-war baby boom in full swing and so many new mouths to feed, livestock began to be managed on an industrial scale and breeders turned to technology to help them make the most from their herds. In arguably the most important

---

* There are also examples of exceptionally buff double-muscled sheep, whippets and people. In 2004, scientists described a boy who was so muscly that aged four, he could extend his arms while holding a 3kg (7lb) dumbbell in each hand.

development in the history of cattle farming, breeders started to collect semen from star bulls and use it to inseminate vast numbers of females. Sexual reproduction, as practised by the birds, the bees and the aurochs too, was usurped by artificial insemination (AI).

It doesn't take much to get a bull excited. Breeders realised that if bulls were presented with the backside of another animal, most will readily mount and ejaculate. It doesn't matter if the other animal is a cow or even a neutered bull. The response is hard-wired. To collect the semen, 'all' the breeder has to do is 'intervene' at the critical moment and divert the erect penis into a suitably shaped receptacle. Now prize bulls didn't have to be physically transported to the females they were due to service. The bulls could stay put, and their semen could be shipped instead.

Today, AI has become a mainstay of agricultural practice, and because semen can be frozen and stored, the seed from a single star animal can be used to inseminate hundreds of thousands of partners. Romance is dead, replaced by a glorified bovine turkey baster. It's used in the swine-, equine- and sheep-breeding communities and the technique is often used to flood the gene pool with the DNA of a preferred, chosen few. As a tool of selective breeding, it enables desirable genes underpinning desirable characteristics to become widespread. Now all breeders – or at least those that can afford it – can have access to successful animals.

Artificial insemination was used to beef up the already buff physique of the Belgian Blue. Since the arrival of AI, it has transformed from being a stocky, well-muscled animal into one of comic-book proportions. If you commissioned Jeff Koons to make a model cow using only a dozen balloons and a pair of nylon tights, this is what you would get: a beast that looks as if it has been inflated rather than allowed to grow naturally.

In the dairy industry, AI has been used to boost milk production. In the US, 60 per cent of dairy cows are artificially inseminated, while in the UK, the figure is even higher. Many breeders don't even bother to keep bulls. They just buy 'straws' of semen online and such is the pace of progress, it's now possible to buy sperm that has been sexed and sorted. If you want to skew the odds in favour of producing a female rather than a male calf, it's now possible to do so.*

The result is that today, the average cow produces four times as much milk as one from the 1960s. Holsteins like Knickers are the most common dairy breed in the UK, Europe and the USA, with cows producing between 20 and 30 litres (35–50 pints) of the white stuff per day. But there's still a problem. In the dairy sector, it easy to decide which females you should use as breeding stock – the ones that produce the most milk. Bulls, obviously, don't produce milk, so how to assess the quality of a future stud that you hope will produce productive daughters?

Prove It

For the last 50 years, the answer has been a method called 'proving' in which the bull literally has to prove

* Male sperm carry a small, runty Y sex chromosome, while female sperm carry a magnificently large X one. This makes it possible to effectively sort the sperm by weight, but the process is not foolproof. Even with the best technology, a sexed semen straw will contain around 90% female and 10% male semen.

himself. A young male is selected and reared to sexual maturity. His semen is then collected and used to inseminate maybe 100 randomly chosen cows, which are then allowed to have calves. When the female calves are around two years old, they are allowed to breed and it's only then that their milk-producing prowess can be assessed. All the while, the original bull has been standing around twiddling his horns. He's now four years old and has cost the farmer around £40,000 (US$50,000). In the US alone, around 12,000 bulls are left in this kind of limbo every year. To top it all, when the process is finished, only about one in ten bulls is judged good enough to be classified as a high-quality stud. It's an expensive, time-consuming and wasteful process.

Then in 2009, researchers decoded the genome of the domestic cow. The genome is the entire set of genetic information found inside a cell or organism. It's made of deoxyribonucleic acid (DNA), that famously twisted molecule that is made up of four simple chemical nucleotides or 'letters' known as adenine (A), cytosine (C), guanine (G) and thymine (T). Genomes are often billions of letters long, so the DNA is divided up into more manageable chunks called chromosomes. The DNA, which is housed inside a membrane-bound structure called a nucleus, acts as a blueprint, providing the genetic instructions that are needed to build an organism. When the cow genome was deciphered, researchers predicted that the information would be of great use to the cattle industry. They imagined being able to pinpoint the genes that underscore commercially useful traits, such as milk production, temperament and longevity, but as time went by, the relevant genes remained elusive. It's no surprise. One of the big lessons we have learned from modern genomics is that complex characteristics, such as personality, life span and indeed milk production, don't follow the neatly defined, compartmentalised Mendelian laws of inheritance. There is no single gene responsible for any of these features. Instead, most complex characteristics are underscored by multiple genes which act together to produce a cumulative effect.

What the scientists did notice, however, are tens of thousands of locations scattered across the cattle genome that often contain single-letter changes to the DNA code. These single-letter changes are known as single nucleotide polymorphisms or SNPs (pronounced *snips*) for short. Today, with the advent of big data and complex number-crunching algorithms, it's been possible to link many of these SNPs to physical characteristics. Researchers aren't saying, for example, that a 'C' being replaced by a 'G' at the far end of chromosome number 2 is *causing* a cow's milk yield to increase, but they are saying they've noticed a correlation, and this in itself is enough to be helpful.

San Diego biotech company Illumina were among the first to market a genetic test that predicts which of many characteristics a cow or bull may develop. The test relies on a SNP chip: a small sliver of plastic about the size of a fingernail, that has more than 50,000 nanoscale dints etched into it. When DNA from an animal is sluiced across the chip, it binds or 'hybridises' with specially designed probes which then 'light up' if a particular SNP is present. From this, SNP chip tests can predict all manner of possible future subtleties, including the quality and quantity of an animal's milk, the length of its productive lifespan, how docile it will be and the age it will be when it has its first calf.

The test has been widely adopted by the US dairy industry, where it is rapidly replacing the old-fashioned practice of proving. There is no longer any need to wait around for the progeny of unproven bulls to grow up and start making milk. Now, if a farmer wants to know how good a potential dairy sire will be, all he needs to do is remove a single tail hair from the newborn calf and send it off to Illumina. The test costs around US$40 (£32) per bull, and the results come back within days. As a result of this and other SNP chip tests, more than half of all AI matings in the US are now made using genomically tested bulls. Ten thousand years after the aurochs was domesticated, 300 years after selective breeding began, the process of artificial selection has moved into a new phase. It has entered the genomic age.

George Seidel is an Emeritus Professor at Colorado State University. He has spent his career developing clever reproductive technologies for the cattle and equine breeding communities, including how to sex semen from bulls. He has watched the global rise of artificial insemination, but when I ask him which technology has most influenced the cattle breeding industry, he is torn. 'Artificial insemination is extremely powerful,' he tells me, 'but SNP chips are changing the industry. Compared with five years ago, things are now being done completely differently. The old paradigm has disappeared.'

Bouffant Bantams

It's worth, at this point, pulling focus away from the bovine world to look more widely at other aspects and products of selective breeding. Most breeds of dog, cat, pigeon, chicken, cattle, sheep and pig are recent inventions. They are products of an intense selective breeding craze that spans the last few centuries. There are now over 800 different breeds of cattle, more than 300 breeds of dog, and more than 40 recognised breeds of domestic cat … and don't even get me started on the crossbreeds.

Annie is our oldest chicken. If you asked a child to draw a chicken, she's the sort of bird you would get; a medium-sized, conker-coloured clucker with scaly feet and a bright red comb. She's a Ranger; a breed selected for its fast growth, prodigious egg laying and tasty drumsticks. Annie is not for the pot, however. In our house, hens are pets so this would never be allowed. Then there's Simon Cowell,* the most recent edition to our chicken family. A slip of a bird, she has fluffy beige plumage, feathery feet with an extra toe, and instead of a comb has a frizzy eighties hairdo. Simon Cowell's head looks like an eggshell that has sprouted cress. When it rains the unfortunate coiffure flops on her face and she finds it hard to see. Although her provenance is uncertain, she's almost certainly part Silkie. Silkie-like birds have been known for some time. Marco Polo

* Simon Cowell is a female chicken, named by my daughter's friend Rosie. We considered changing her name to Simone Fowl, but were overruled by the children.

wrote of 'furry chickens' back in the thirteenth century, but the breed wasn't formally recognised until 1874. As its popularity grew, breeders hoodwinked buyers by telling them that the unusual birds were the offspring of chickens and rabbits, while sideshows promoted them as having actual mammalian fur. Somewhere along the line, we got carried away and stopped selecting animals for pragmatic qualities, like their meat or ability to churn out eggs, and started to select them for their aesthetics. Simon is small and bony. Her extra toe is of no practical use. It flops around uselessly at the base of her foot. Nor do her unusual downy feathers give her a survival edge; but she is beautiful … if you like your bantams bouffant.

We also have two rabbits, Ella and Brownie. Like our chickens, they are the products of selective breeding. Ella has a white coat with black ears and is of uncertain heritage, while Brownie, who has a shaggy brown mane, is at least part Lionhead. It's thought that Lionheads originated in Belgium when breeders were crossing different bunny varieties together. At some point in time a mutation cropped up that causes a woolly ruff of fur to grow around the neck. People liked the look of it, so the feature was selected for.

Sometimes we have selected, not for looks, but for quirky behaviour. As pigeons go, the Birmingham roller is nothing much to look at. It's a cobalt coo-er that would fail to turn heads in London's Trafalgar Square, but watch it fly – or try to – and the animal is quite remarkable. Adult roller pigeons take to the air as normal, and then without warning begin to tumble from the sky.

They perform a series of rapid backward somersaults. There's something almost balletic about it. The birds tumble repeatedly in apparent freefall before regaining their composure and avoiding impact with the ground.

Then there's the myotonic goat, a strain of domestic goat that freezes and falls over whenever it is spooked. Animals go as stiff as a board, wobble momentarily, then capsize onto their back with their legs sticking up in the air. The whole episode can last up to 20 seconds, during which time the immobile bleater can be picked up like a plank of wood. After that, they are absolutely fine. The breed can be traced back to the 1800s, when a farm labourer called John Tinsley moved to Tennessee and brought the animals with him. Locals liked them because, unlike most other varieties of goat, they didn't climb or jump much. This made them easy to fence in, and if the animals ever did make a bid for freedom, they'd inevitably get skittish, have a fit and fall over. Their unusual habit earned them various monikers including the fainting, nervous, stiff-leg and wooden-leg goat, but the response is less of a faint and more of a spasm. The animals carry a single letter change to a gene which normally helps control the flexing and relaxing of muscles. Tensing up is part of the normal fight or flight response, but a mutated version of the gene means that the muscles don't relax as quickly as they should. In the 1980s, farmers took a fancy to the myotonic goat and began to breed the smallest and stiffest animals together, for pure novelty value. The result is a breed whose behaviour is so quirky it has racked up hundreds of thousands of YouTube hits. Google 'fainting goats'. I urge you.

We have, at times, been shallow and vacuous with our breeding programmes, driven by fads and novelty. Many different species have been selectively bred on what is little more than a personal whim. Budgies and canaries have been bred in just about every colour of the rainbow while dogs have been shrunk so small they can fit inside a handbag. We have made bald cats, long-haired hamsters and goldfish with Elvis-like quiffs. Even wild animals are not exempt from this wanton evolutionary mischief-making.

Golden Gnus

In South Africa, the owners of commercial game reserves have started selectively breeding wild animals, such as wildebeest (or gnu), impala, kudu and springbok. Occasionally mutations crop up that change the animals' colouring. Wildebeest and kudu are normally a dark, grey-brown colour, while impala and springbok are a sandy beige. Now there are golden gnu, black springbok and white kudu. Impala now come in white and black forms, as well as a 'split' variety whose upper and lower body is divided into different colours. In natural circumstances, these animals would be unlikely to survive long, or pass on their unusual genes to future generations. Sticking out like a sore thumb is not a brilliant survival strategy. Unless you're at the very top of a food chain and can afford to be brazen, blending in tends to be a very wise decision. In South Africa, however, these naturally occurring colour mutants are increasing in number as they are deliberately bred in game reserves to supply the trophy hunting market. People actually pay to shoot these unusually coloured wild animals.

In 2015, a 13-year-old lion called Cecil was killed in Zimbabwe's Hwange National Park. With his instantly recognisable black mane, Cecil was a firm favourite with tourists and part of a scientific study by researchers from the University of Oxford. He met his end when an American dentist shot him with an arrow, then tracked the wounded animal for 40 hours before finally killing him with a second shot. The lifeless beast was then skinned and beheaded.

The episode sparked worldwide controversy and heated debate over the ethics of so-called trophy hunting, but what became obvious was just how unaware most people are about the reality of modern-day African hunting. In Africa, big game is big business. The practice generates an estimated £160 million (US$200 million) per year for the African economy, and over the last 40 years there has been a shift away from pastoral cattle-ranching to big-game reserves. Now there are around 12,000 fenced-off game parks, where tourists can go either to photograph Africa's unique evolutionary heritage or to kill it … and/or eat it. In South Africa, it's legal to own wildlife, sell wildlife, breed wildlife and kill wildlife. Lions, elephants, rhino, buffalo and gnu are just a few of the species on offer.

Just for a moment, put aside any visceral reaction that you may have to this information. I find the hunting of wild animals for pleasure repugnant. I think it's warped that people should selectively breed unusually coloured wild animals just so they can be killed, but I have to acknowledge this is a complicated situation. On the one hand, this decentralised approach to conservation – where the management of wildlife is determined by the practices of individual ranch owners – can be detrimental to wildlife. Game ranchers keep their high-value animals, like impala and springbok, in massive enclosures surrounded by supposedly predator-proof fences, but the fences aren't perfect. Sometimes captive-bred ranch animals come into contact with free-roaming wildlife, such as lions, cheetahs, jackals and hyenas. The result is bloodshed; first when the predator kills the rancher's property, and second when the predator is shot dead. Studies have documented a growing intolerance towards these so-called nuisance animals, and critics warn that their ongoing persecution is contributing to the loss of ecologically important species. On the other hand, South Africa is widely heralded as a conservation success story. Research has shown that since the country moved away from cattle farming and embraced big-game reserves, overall, wildlife has boomed. There has been a significant increase in the abundance and distribution of many wildlife species. The money earned from ecotourism and trophy hunting can be ploughed back into the reserves to the benefit of biodiversity. Although it may seem counter-intuitive, sometimes, hunting can actually help wildlife.

For the golden gnu and all of its mutant associates, this is unlikely to be the case. Their selective breeding began over 10 years ago, when naturally occurring colour variants were captured from the wild and brought into fenced game reserves. The owners of these reserves thought that if they could persuade the animals to breed, trophy hunters would pay handsomely to shoot the descendants. Early signs were promising. In 2012, colour-variant animals were being actively traded between breeders as they all tried to establish their own stock. Where a regular impala would sell for around R1,400 (£80 or US$100), a black one would fetch more than 400 times that amount: R600,000 (£34,000 or US$40,000). Two years later, as the number of colour variants increased, the average price for white impala rams reached a staggering R8.2 million (£460,000 or US$550,000). Bear in mind that at this point, no one had actually paid to shoot one of these animals. These were just individuals that were being traded between reserves to establish speculative stock. Colour variants had become ridiculously expensive because they were rare and because breeders expected that when they entered the wider market, they would be in demand. How wrong they were.

When the animals became available for shooting in 2016, uptake was slow. There wasn't anything like the anticipated excitement. Since then, their value has plummeted. The price of a black impala has fallen to around R10,000 (£560 or US$700), while white impala sell for around R48,000 (£2,700 or US$3,000). Critics pointed out the dangers of breeding intensively from limited founding numbers and cautioned that inbreeding could become a problem. Some dubbed the animals 'unethical' because they were captive-bred, while others viewed them as unnatural, undesirable freaks of nature.

The bubble burst. Hunters simply didn't want to pay the exorbitant fees being charged to kill these designer animals. The breeders got it wrong. The expected market demand did not exist. Now an estimated 5 per cent of South African game ranches own these unusually coloured animals and no one is really sure what their fate will be. Ranchers selectively bred on an aesthetic whim, to what end? A game park stocked with golden gnu that few people want to hunt.

Meanwhile, ranchers continue to selectively breed for what they believe will be the 'next big thing'. In the last couple of years, breeders have begun to advertise captive-bred buffalo, sable and roan.* These animals have normal coloured hides but their horns are enormous, and as a result, they have become valuable trophy animals. In 2016, a single big-horned buffalo bull sold for R168 million (£9.5 million or US$11 million). It's all very well to selectively breed for a single particular trait, but the repercussions can be unpredictable. Domestication taught us that. Remember Belyaev's silver foxes, selected solely for tameness? It wasn't just their behaviour that changed. They also changed physically and physiologically. The long-term consequences of selecting solely for horn size are as yet unknown.

Selectively Breeding Trouble

As selective breeding nudges some wild animals towards an uncertain future, the practice is also creating problems for domesticated animals. This is most evident in the dog world, where in recent times breeders have exaggerated relatively benign features and made them pathological. Before the 1835 Cruelty to Animals Act made it illegal in the UK, Bulldogs were put to work in bull-baiting fights. In Queen Anne's time (the early 1700s) they were pitted against tethered bulls which the dogs had to pin to the

---

* Sable and roan are both types of antelope.

ground. A successful Bulldog was agile, thickset and muscly. Its broad head had a slightly undershot jaw so it could grip onto its opponent's face, and a nose set back behind the jaws so it could breathe while holding on. After bull-baiting came to an end, the English Bulldog breed society was set up in the late 1800s and slowly but surely, the physical idiosyncrasies became magnified. Compare the skull of a modern Bulldog with that of its bull-baiting counterpart. They look like different species. The modern Bulldog has a shortened snout, flattened face and jutting lower jaw. Its head is thickset and the exterior is cowled in folds of hanging flesh, all of which conspire to make the breed prone to overheating and breathing difficulties. Sometimes these problems necessitate surgery, and the average Bulldog puppy's head is now so broad that it struggles to fit through the birthing canal. Some 80 per cent of Bulldog pups are now born by Caesarean section. In a similar vein, Dachshunds are prone to back problems, Basset Hounds suffer from ear infections, and Pugs' bulging eyes sometimes pop out of their sockets because their skulls have become too small to house them.

In the agricultural sector, things are far worse. Over 90 years ago, it took a newly hatched broiler chick at least 16 weeks to become oven-ready, but now selective breeding has produced birds that reach slaughter weight in just over 4 weeks. During the same 90-year period, their market weight has more than doubled from 1.2kg (2.6lb) to 2.5kg (5.5lb). This accelerated growth rate takes its toll on their hearts, lungs and limbs, which are unable to support their over-sized bodies. The birds look like weightlifters on steroids. 'Most live in factory farms where they're often in crippling pain and unable to walk,' says Philip Lymbery, Director of Compassion in World Farming, 'to the point where some just die because they can't make their way to the food and water.' Deaths from heart attacks or from swollen hearts unable to pump enough oxygen around their oversized bodies are also common. Things are little better for egg-laying hens, which have been selectively bred to produce hundreds of eggs per year. This depletes their calcium reserves and results in high levels of osteoporosis and fractures.

These examples are just the tip of an iceberg littered with centuries of short-sighted decision-making. Our pursuit of aesthetic and commercial ideals has created a selective animal welfare blind spot. We have selectively bred animals that either cannot survive on their own, or suffer because of the characteristics we have selected. Were these wild animals, natural selection would have weeded them out decades ago. We selectively breed at our peril and often the most serious repercussions are the ones we can't physically see.

In the 1970s there was an American Quarter Horse called Impressive. Quarter Horses compete in short races of a quarter-mile or less, so the best individuals are quick off the mark and incredibly fast. Their sprinting power comes from having well-muscled hind quarters, so over the years, Quarter Horse breeders deliberately selected for this feature. Impressive had a refined and muscular form, and a winning disposition. After he became a World Champion in 1974, every Quarter Horse breeder wanted a piece of his genetic legacy. He was put to work as a stud and via artificial insemination, went on to sire more than 2,000 foals, which then grew up and had foals of their own.

So far, so good, but in the 1980s, a problem emerged. Although Impressive was impressive, some of his descendants were not. Some had developed a strange muscular twitching that caused them to fall over or left them immobile. Some horses even died. Because only Impressive's descendants were affected, and the syndrome had never been seen in other horse lineages, a genetic cause was suspected.

A gene called '*hypp*' was implicated. Unbeknown to anyone, Impressive carried a mutated version of this gene, which he had passed on to some of his descendants. On the surface, the mutation was beneficial. It alters the electrical activity of muscle cells, endowing its carriers with their fine, muscly frame, but if two copies rather than one are inherited, the cellular effects are exacerbated. This can be deadly. The gene that helped dozens of Impressive's descendants become World Champions was the same gene that harmed and killed so many more. It's estimated there are now more than 300,000 living descendants of Impressive, and that at least 4 per cent of all Quarter Horses still carry this mutated gene.

Returning to cattle, there are similar horror stories. In 1962, a Holstein bull called Pawnee Farm Arlinda Chief – or 'Chief' for short – was born on a US farm. He had excellent genes for making milk and went on to become one of the most important sires in the Holstein breed. Through artificial insemination, he had 16,000 daughters, 500,000 granddaughters and over 2 million great-granddaughters. His sons were also popular sires and as a result, Chief's genes spread widely through the Holstein breed.

Unlike Impressive, who had poorly descendants, Chief's family was in fine health. No one realised anything was amiss until researchers studying SNP chip data noticed something unusual. SNP chips, you'll remember, work by association. By gathering together vast amounts of data, researchers have linked single letter changes in the DNA code with overt features such as milk production. Scientists noticed an association between one set of SNPs and an increase in the rate of stillbirths in Holstein cows. All they knew about these SNPs was that they resided somewhere on chromosome 5, and that this unusual genetic signature could be traced all the way back to Chief.

When Chief's genome was sequenced, the culprit was found. The SNPs fell within a gene called *Apaf1* that had been well studied in mice. Mice who inherit two faulty copies of the gene die while they are embryos, and a similar thing had been happening in Holsteins. Chief's living family *were* in good health, but no one had really paid much attention to the unknown descendants that were miscarried. A mutation in Chief's *Apaf1* gene that proved lethal when two copies were inherited had unwittingly been spread through his family. Those animals with a double dose were never born alive.

It's estimated the mutation has caused half a million spontaneous abortions in Holstein cattle over the last 35 years, accounting for around US$420 million (£335 million) in losses. All this, because of a single faulty gene that was inadvertently spread through a population when breeders were selecting for milk production.

What is perhaps most baffling is that many of these dangerous mutations still exist, now that we know about them and have developed genetic tests to spot them. Katrin Hinrichs and colleagues at Texas A&M College of Veterinary Medicine developed a test for the mutated *hypp* gene that can be used on embryos. This means that affected horses need never be born. A similar screening procedure exists for Holstein cattle. The tests are like human Preimplantation Genetic Diagnosis (PGD), a technique that is used to screen IVF embryos for certain genetic defects, to prevent genetic diseases from being inherited.

Genetic tests like this provide a way to stop harmful mutations from spreading. 'You could eliminate these mutations in one generation,' says Katrin, yet when she proposed her test to the breeding community, they were surprisingly uninterested. It turns out that horses carrying one copy of the mutated *hypp* gene are just too good to give up. People want to keep breeding from Impressive's descendants because the family genes offer a competitive edge. Meanwhile, although Holstein breeders are beginning to adopt the genetic test for Chief's faulty gene, they too can look past this and see the broader economic picture. Although the faulty gene cost the dairy industry millions of dollars in losses, using Chief's sperm to inseminate dairy cows has still led to US$30 billion (£24 billion) in increased milk production over the past 35 years.

We have arrived at a situation where decades of selective breeding, guided by the technology we have developed, is creating problems. In the cattle sector (and elsewhere) as breeds became established, populations of cows became physically isolated from one another and reproduction between these groups was reduced. In terms of the broader genomic landscape, the genetic vibrancy that comes from the interbreeding of healthy, non-related animals began to be eroded. Highly productive breeds have been intensively selected for particular traits, with little emphasis placed on the preservation of genetic diversity. Artificial insemination has become an important breeding tool, but it inevitably reduces the number of genetic variants that are passed on to following generations. Because a single sire can father tens of thousands of offspring, there is now more genetic 'sameness' both within and across generations. This is not good. Now many breeds are suffering the effects of inbreeding. This means they have an increased risk of genetic disease and fertility problems, and as populations, may be less able to fight off new, emerging infectious diseases. We are now beholden to an industry that prioritises short-term productivity over the long-term health and sustainability of its breeds.

If cattle were wild animals, we would look at them and see them for what they are: fragmented populations with dwindling genetic diversity. We might even consider them to be endangered. The development of industrial breeds, like Holsteins and Aberdeen Angus, put pressure on farmers to abandon many traditional breeds in favour of these cash cows. As a result, 100 traditional livestock breeds have gone extinct, and a further 1,500 are deemed at risk of the same fate. This potentially puts our food security in jeopardy, and geneticists are now genuinely concerned for the future of cattle.

A Load of Old Bull?

As the number of domestic cattle rose, the number of aurochs fell. Human hunting nudged the aurochs further along their collision course with extinction. Their range dwindled until 800 years ago, aurochs could only be found in parts of eastern Europe. As they became rarer still, the right to hunt them became restricted to royalty, but this was too little, too late. In 1627, the last ever aurochs, a female, died in the Jaktorów Forest in Poland.

In 2015, researchers decoded the aurochs' genome. DNA prised from a bone found in a Derbyshire cave revealed something that had been expected, but never conclusively proved. At some point in the past, wild aurochs and early domestic cattle interbred. In a sense, the aurochs never died out. Vast swathes of its DNA still persist in living breeds of cattle.

This genetic legacy is most apparent in traditional UK cattle breeds, such as the Highland, the Dexter and the Welsh Black. These same breeds contain other unique genetic variants not found in the mass-produced, genetically homogenous varieties that we rely on so heavily. Crossbreeding between them could help to raise those all-important levels of genetic diversity, and enticingly, it could also do something else.

If the genes of the aurochs live on in its modern-day descendants, it should be possible, through selective breeding, to bring the aurochs – or something that approaches it – back. This is not a new idea. In the nineteenth century, zoologist Feliks Pawel Jarocki described how it might be possible to do just this using back-breeding, where breeders take cattle with aurochs-like features and then select for these traits over time.

In the 1920s, German zoo directors Lutz and Heinz Heck gave it a go when they started crossing various breeds of European cattle. The result, in the 1930s, was a superficially aurochs-like breed known as Heck cattle. With the rise of the Nazi party and backing from Hitler's right-hand man, Hermann Göring, the cattle played into an unpalatable narrative of ethnic cleansing and eugenics. The animals were meant to show that selective breeding could be used to 'improve' a population. With their muscular frame and powerful horns, the beasts were used as an allegory for the Nazi party's strength, and Göring had them released into his private hunting grounds. When Allied forces closed in on the Nazis at the end of the Second World War, many of the Heck cattle were killed but a few lived on. Today their descendants can still be found in zoos, nature reserves and a few farms.

Although Heck cattle are physically on point, they still lack many of the aurochs' finer features such as its lean limbs and long skull, so since then, others have tried to recreate this iconic beast. Today, the rationale is reassuringly different. Attempts to recreate the aurochs have an ecological rather than an ideological basis.

In times gone by, big grazers, like the aurochs, provided a natural gardening service. By grazing open spaces and fertilising the land, they created the conditions for other species to thrive. Now Europe is missing a large, predator-proof herbivore and researchers believe that if herds of aurochs-like animals were returned to the wild, they could work their magic and boost biodiversity levels.

Ronald Goderie is founder of the Tauros Programme, a project seeking to create an animal that not only looks like, but critically, acts like an aurochs, so that it can help to 'rewild' vast swathes of central Europe. Focusing on physical characteristics, behaviour and genetics, scientists with the Programme have spent over a decade breeding together various primitive, hardy cattle breeds such as the Spanish Pajuna and the Italian Maremmana. Now, four generations later, the animals are increasingly looking the part. The bulls are big and black, with long heads and impressive, thick horns. The bulk of their mass is concentrated over their front rather than their hind legs, and the tail hangs freely just like a wild bovine. Females, in contrast, are smaller, thinner and sport a coat in dark earthy, ochre tones. In the early days, artificial insemination was used to accelerate the breeding process, but now as the animals settle into their wild test sites, things are becoming more hands-off. In 2013, Ronald teamed up with Rewilding Europe, a non-profit-making conservation group that aims to rewild 1 million hectares of European land. There are now herds of Ronald's Tauros cattle in Spain, Portugal, Croatia, Romania, Belgium and the Czech Republic.

Enticingly, features of the aurochs' naturally wild behaviour are already evident. 'It astonished me,' says Ronald. 'You see this even with the first generation of animals. The calves are small and easily born. We don't get involved at all. Then sometimes you think the calf has died because you can't find it, but what's happened is that the mother has hidden it.' She visits a couple of times a day to offer milk and may keep her offspring hidden for weeks. This is what wild bovines do to protect their young from predators. Another wild behaviour is the establishment of bovine kindergartens. 'All the calves are kept together with one or two cows keeping an eye on them,' explains Ronald, 'then they rotate their roles.' Crucially, Tauros cattle released into the forested canyons of Croatia's Velebit Mountains have already proved their credentials. Native herbivore populations are regularly decimated by wolves, but the Tauros cattle seem able to cope. Just like the original aurochs, the Tauros can defend themselves against wolves and there have been few losses to predators.

In the meantime, selective breeding continues and there's no reason why, as technology improves, geneticists couldn't create a SNP chip full of aurochs-like features to help guide future matings. By 2025, Ronald and his team envisage self-sufficient herds of at least 150 animals in several rewilding sites across Europe. They won't *be* aurochs, but they will look and act like aurochs. Their grazing services will be a boon to the local landscape, and their presence could offer a genetic lifeline to their increasingly beleaguered descendants.

CHAPTER THREE

Super Salmon and Spider-Goats

In June 2017, Canadian consumers picked up their chopsticks and tucked into some delicious, high-end sashimi. The delicate pink cuts of flesh were neatly arranged on a bed of shredded daikon radish, and served with a dab of wasabi and a sprinkling of soy sauce. They slipped down a treat, and the consumers thought little more about it until, a short while later, a fishy story emerged.

In August of the same year, an American company announced that it had sold 4.5 tonnes of genetically modified (GM) Atlantic salmon to unnamed businesses in Canada. The fish didn't look or taste any different to regular salmon, and in Canada, supermarkets don't have to label GM ingredients on their packaging, so the fish swam under the public's radar and right onto their dinner plates. The AquAdvantage salmon, as it is called, became the first genetically modified animal to become part of the human food supply.

When the story was reported, reactions were mixed. While some went back for a second helping, others were annoyed they had bought the product without knowing its provenance. They lamented the lack of mandatory labelling and bemoaned the perceived absence of transparency that had brought the fish to their table. Environmentalists were worried that the 'Frankenfish' might escape and wreak havoc in the natural world, while activists demanded that the regulators do a U-turn and put the slippery genie back in the bottle. Welcome to the divisive world of GM organisms where, it seems, there's never a dull day.

After domestication and selective breeding, GM is the next significant milestone in our mastery of the evolutionary process. Genetic modification is the process of altering an organism's genetic makeup. Today, much of the food we eat has been genetically modified in some way or other. If GM fish sound unsavoury, nevertheless, at some point in your life, you've more than likely consumed GM products. GM plants are commonplace in our supermarkets and on our plates. In 2017, 24 countries planted a combined 1.9 million square kilometres (470 acres) of GM crops. That's equivalent to an area the size of Mexico. A leader in the field, the United States cultivated an area the size of Turkey, and more than 90 per cent of the country's maize, soybean and canola is of genetically modified origin. Around the world, farmers routinely plant GM versions of sugar beet, alfalfa, papaya, squash, aubergine and apples. We have GM spuds that bruise less easily, GM tomatoes that stay ripe for months and GM rice strains modified to be more nutritious, productive and disease resistant. Hundreds of millions of people have consumed these crops without incident, and with the AquAdvantage salmon now swimming onto the scene, the stage is set for an influx of GM animals. Dozens of fish species are having their DNA deliberately altered, as are cows, chickens, pigs, goats and other farm animals.

On a wider scale, this is just the tip of the genetically modified iceberg. As increasingly sophisticated methods emerge for tinkering with DNA, scientists are deliberately redesigning the genomes of living things with implications far beyond the realms of food production. Animals are being altered to produce novel medicines and materials. There are goats that make spider proteins in their milk, and chickens that produce drugs in their eggs. Pigs are being modified so their organs can be transplanted into people. GM is a broad continuum. On the one hand, it is being used to make designer pets like 'glow in the dark' fish; on the other, it is being developed to help understand and thwart disease. At one extreme, the technology is being used to destroy life forms we deem undesirable, while at the other, it's being used to bring extinct species back from the dead (see Chapter 4). From the flippant and the frivolous to the profound and the planet-changing, GM is one of the most powerful technologies ever developed.

Humans have been genetically modifying things indirectly for thousands of years as they selectively bred their animals and crops. Dogs really are genetically modified wolves. The tiny 0.5 per cent that is different is sufficient to separate the wolves from the whippets, and afford dogs their subspecies* status within the canid

---

* Life is categorised into different ranks, ranging from domains and kingdoms all the way down to the familiar families and species. Subspecies are a rank below species.

family. Wolves are *Canis lupus.* Dogs are *Canis lupus familiaris.* In recent times, however, we've assumed a more deliberate mode of control. When we fashioned dogs from wolves, or chickens from red junglefowl, we didn't consciously alter the animals' DNA; rather we selected features that we liked, and the DNA patterns that underpinned them hitched along as part of the ride. Then, a century or so ago, things started to change when we started to deliberately hone in on DNA.

The Red Canary

In the 1920s, a German schoolteacher called Hans Duncker decided to go after the genes that give some birds their vibrant red feathers. An avid bird breeder, he was entranced by the many varieties of canary that existed. Over centuries of selective breeding, fanciers had remodelled the naturally green wild canary into various shades of white, yellow and orange, but no one had ever managed to produce a canary with red feathers. Duncker was determined to try, and reasoned the only way he could achieve this was to take the 'red genes' from another bird species – the red siskin – and add them into canaries.

Today this would be relatively easy. With modern techniques it's now fairly straightforward to add new genes into old genomes, and mix and match the DNA of different species, but 90 years ago this was groundbreaking territory. The only tools he had were the theories of Mendel and Darwin, and the hard-won knowledge of enthusiastic bird fanciers who knew seemingly everything about persuading caged songbirds to breed.

Red siskins are small, vibrantly coloured finches from South America. When Duncker was planning his experiments, wild red siskins were already being shipped to Europe as part of the captive bird trade. His plan was to use a clever breeding paradigm that would see the canary genome first flooded with red siskin DNA, then refined so that only a trace of the siskin's DNA – the genes that code for red feathers – remained.

The plan came in three parts. Stage One was to mate a male red siskin with a yellow canary to produce a hybrid that was half-siskin, half-canary. In the past, when people had crossed the two species, the offspring tended to adopt the colour of the finch rather than that of the canary. So Duncker hoped his hybrids – also known as 'mules'* or 'bastards' – would have red feathers. Stage Two was to breed these mules together to create 'doppelbastards'. According to Mendelian principles, he expected this second generation of birds to sport red plumage in varying shades. Some of the birds would carry a double dose of the red genes and so be even more carmine than their parents. Stage Three, the final step, was to cross these vibrant red mules back to a canary, and for Duncker to keep his fingers crossed. This last step was designed to concentrate the genes for redness and whittle away all the other unneeded siskin genes.

* In the birding world, the term 'mule' has a very specific meaning. In Duncker's time it was used to denote a cross between a canary and a red siskin, as opposed to a cross between a canary and any other kind of finch.

After four or five years, Duncker hoped to produce a bird that was predominantly canary, but with a spattering of choice red siskin genes – the world's first transgenic animal.

Transgenic organisms are living things imbued with small, choice amounts of DNA that comes from different species. The sweet potato, for example, is a naturally occurring transgenic plant which contains genes from a single-celled organism, *Agrobacterium*. It wasn't engineered in a laboratory or produced by some clever breeding method; rather, 8,000 years ago the plant was infected by the bacterium, then somehow retained a fraction of the microbe's genes. The AquAdvantage salmon described at the start of this chapter is not only genetically modified; it is also transgenic because it contains choice DNA sequences from two other fish species. In contrast, hybrids, which sometimes occur when different species interbreed and have offspring, contain similar amounts of DNA from both parent species. They are different from transgenic organisms because they tend to be half one thing, and half another, so lack the nuanced, deliberately engineered genetic makeup of a truly transgenic organism.

Duncker began his quest for the transgenic red canary in the spring of 1926. Stage One was a breeze, but Stage Two was a problem. After he had successfully crossed a red siskin with a yellow canary to produce coppery mules, the mules did not want to breed with each other. They skirted around each other like awkward teenagers at a school disco and the end result was just a few, infertile eggs. When Duncker dissected the birds, he found the root of the problem. The frigid females had no internal reproductive organs. Small wonder they had little interest in sex; the cross-breeding process had effectively given them a hysterectomy. Reassuringly, the gonads of his male mules were normal, so Duncker decided to skip Stage Two, and move straight on to Stage Three. The coppery male mules were crossed back to a yellow canary hen. Duncker's earlier experiments had suggested that in canaries, colour is inherited in a classically Mendelian manner and that yellow is a recessive characteristic. So when he mated a mule with a canary, he expected that three-quarters of the offspring would be reddish, and a quarter would be yellow.

If only it had been that simple. Although the birds produced offspring, when they grew up their feathers did not obey Duncker's neatly predicted ratio. Instead, all of the birds bore rusty-coloured feathers, just like their fathers. Although these birds definitely did not look like a standard canary, they didn't have the vibrancy of the red siskin either. Duncker concluded that the canary's yellow colouring was somehow masking the appearance of siskin's scarlet hues, so he tried crossing his coppery mules with white canaries, but again no joy. The offspring were either copper-coloured or grey.

After that Duncker bowed out of the quest for the red canary, and it took an international cast of bird breeders to finish the job. In Prussia, a man called Bruno Matern repeated part of Duncker's research, and crossed one of the russet-coloured mules with a yellow canary hen. This time, bizarrely, the results were different. Instead of replicating Duncker's copper-coloured offspring, Matern's experiment yielded birds that were orange. In England, a British fancier called Anthony Gill took some of these orange offspring and bred them back to another coppery male mule. The resultant birds were an even deeper shade of tangerine.

The final piece of the puzzle was solved by Charles Bennett, a physiologist from the University of California at Berkeley, who realised that genetics alone was unlikely to produce a red canary. He noted how sometimes, the vibrant colour of imported red siskins faded when they were kept in captivity and reasoned something must be missing from the birds' diet. Bennett knew of a condition called 'carotenemia' that had been described after four women inexplicably devoured 2kg (4.4lb) of raw carrots every week for seven months and turned as orange as a packet of Cheesy Wotsits. In conjunction with this, he referenced a book from 1893 called *The Evolution of Colors in North American Land Birds* in which the author explained how red and yellow were not the discrete entities that Duncker had presumed. Instead they were shades of the same hue that existed on a continuum. The two colours did not compete with one another for expression. Instead, if a bird had yellow or orange feathers, sometimes the potential was there for it to have red feathers as well. Bennett connected this information with his knowledge of carotenemia, and decided to feed some of the orange canaries carrots. In the end, it was simple as that. After the bird had moulted, the feathers that re-grew were a rich and vibrant fiery red. The red canary had finally arrived.

Bennett had discerned that both the genetic makeup *and* the diet of the canary were important. Feeding carrots to a yellow canary had no effect because the birds don't have the genetic predisposition to react to them.

Transgenic canaries containing DNA from the red siskin, however, can be induced to become red simply by feeding them carrots.

In 2016, geneticists were able to pinpoint the exact gene that was at play when they compared the genomes of red siskins and red, green and yellow canaries. The gene, called *CYP2J19*, is thought to encode an enzyme called ketolase that converts carotenoid compounds (which are found in carrots and other substances) into red chemicals called ketocarotenoids. Red siskins carry a mutated version which makes the gene more active, massively increasing the production of ketocarotenoids in the feather follicle cells. This could then be bred into the red canary.

The red canary or 'red factor canary', to give it its proper name, is both GM and transgenic, deliberately engineered via a gloriously low-tech approach that included fancy breeding and carrots. But it was only made possible because the mutation that causes this intense red colouration was already out there. It's a similar story for many other characteristics that are genetically determined, like the mutations that lead to double muscling and the polled trait in cattle. A mutation crops up. People like the characteristic that it codes for and they inadvertently select for it when they breed key organisms together. But the story of GM really ramped up when people discovered ways to deliberately introduce mutations into the organisms they were interested in.

Atomic Gardening

In the baby boom following the Second World War, scientists were looking for ways to improve crops, but they were frustrated by the lack of useful mutations that were already out there. So they came up with a method to vastly expand the repertoire of genetic variation from which they could draw: atomic gardening.

The movement was a by-product of Atoms for Peace, a programme that sought to develop peaceful uses for atomic fission. Scientists knew that radiation had the potential to damage DNA, and could also be used as a way to generate new, potentially useful mutations, so they set up 'gamma gardens'. The gardens were large circular plots, typically covering an area the size of two football pitches. Different plants were laid out like pizza slices, and then a retractable source of gamma radiation, like cobalt-60, was put in the middle. After 24 hours the radioactive source was remotely lowered into a lead-lined box and then scientists, wearing protective equipment, would enter the garden and assess the results.

Seedlings closest to the centre typically died, while those a little further out tended to grow poorly and develop abnormal growths. Beyond these, however, plants in the 'Goldilocks zone' contained mutations but seemed to be growing normally. So scientists let these little seedlings grow and then selectively bred from the best.

Gamma gardens were set up in laboratories in the US, Europe, parts of the former USSR, India and Japan. In the 1960s, the public joined in on the act after an English woman called Muriel Howorth set up the Atomic Gardening Society and started posting irradiated seeds to her members. In one of the earliest crowd-sourced experiments ever, around 1,000 people planted the seeds, then sent back reports about their growth. Although this might sound crazy to us now, given what we know about the dangers of radiation, at the time, these experiments were widely extolled and 'Atom Blasted Seeds' were advertised in shops, magazines and garden fairs. They were even sold at fundraisers by school science clubs.

In the end, atomic gardening fell out of favour as nuclear weapons proliferated and the public became increasingly sceptical about atomic energy. However, we have the process – known as mutagenesis – to thank for more than 2,000 new plant varieties. These include strains that have increased herbicide resistance and drought tolerance, and have enhanced yields and nutritional profiles. The 'Todd's Mitcham' mint cultivar, for example, can fend off the Verticillium fungus and is now widely used to make the mint oil found in many chewing gums and toothpastes, while Golden Promise barley can tolerate high levels of salt and is used to make beer and whisky. Mutagenesis has produced commercially useful strains of sesame, soybean, sorghum, sunflower and sweet potato, and those are just the plants that start with an 'S'.

Scientists had devised a way to create new mutations using radioactivity. This was a huge step forwards, but working with radiation was no picnic and the genetic changes caused by mutagenesis were still random and unpredictable. Mutagenesis like this was a blunderbuss approach to GM and there was no way the method could ever be rolled out to animals because of the collateral damage it would cause. Scientists dreamed of being able to alter specific genes, but even when the structure of DNA was decoded in 1953, the capacity to genetically modify organisms with forethought and direction remained elusive.

Things changed a decade or so later, when scientists developed tools that enabled them to alter genes more precisely by cutting and pasting different bits of DNA together. Rudolf Jaenisch, now at the Massachusetts Institute of Technology, developed a way of adding foreign DNA into mice. He injected viral DNA into embryonic mice and found that it became incorporated into the animals' genetic code. When the embryos grew up, the foreign DNA could be found in every single cell of the animals' bodies.

A little while later, Richard Palmiter from the Howard Hughes Medical Institute, Washington, took things to a new level when he added a functional rat gene into a mouse. He fused a growth hormone gene from a rat with a mouse 'promoter' – a sliver of DNA that helps switch a gene on – and then injected the hybrid DNA into developing mouse embryos. The results, blazoned across the front cover of *Nature* magazine, were sensational. The genetically modified mice still looked like normal lab mice. They had twitchy whiskers, white fur and long tails, but they grew to twice the size. The rat gene caused the mice to keep on growing until they dwarfed the regular lab mice. This caused much hilarity, prompting speculation that scientists should now engineer bigger cats … or larger mousetraps. More than 30 years after bird fanciers used selective breeding (and carrots) to create the world's first non-natural transgenic animal, the red factory canary, Palmiter and colleagues used the tools of modern genetics to do something altogether more sophisticated. They added a specific rat gene into a murine recipient and created the world's first ever transgenic mammal.

Palmiter's goal was not idle curiosity, or a desire to give murophobes sleepless nights; rather he realised that moving genes between species offered a way to study the function of genes, and to generate animal models of disease. Forty years on, and although they can now be made by various different methods, transgenic animals have become a stalwart of the research community, shedding light on the origins of disease and aiding in the development of new treatments, but they are being used in other areas of research too.

A Fish Out of Water

Just like Palmiter's mouse, the AquAdvantage salmon contains the growth hormone gene of one species mixed in with the genome of another. Back in the 1980s, scientists fused a growth hormone-regulating gene from the Pacific Chinook salmon with a promoter from an eel-like fish called the ocean pout. When the hybrid DNA was injected into fertilised Atlantic salmon eggs, it bedded down into the resident genetic material where it caused the fish to grow rapidly. When these fish had offspring, they passed the hybrid gene on to their descendants so they were speedy growers too.

Today's AquAdvantage salmon are all descendants of those first fish. They look like just like any other Atlantic salmon, but grow twice as fast. At 18 months old, when they are ready for market, the transgenic salmon are around 60cm (24in) long and tip the scales at over 3kg (6.6lb). It takes regular Atlantic salmon an additional year to reach this size. Critically, the AquAdvantage salmon eat 25 per cent less feed, and are about 20 per cent more efficient at converting this feed into flesh on their bodies. This means that more salmon can be produced in a shorter time, and the company that now makes and markets the fish – AquaBounty Technologies from Maynard, Massachusetts – think it could be used to help meet the global demand for healthy animal protein. Not everyone is as enthusiastic though.

When AquaBounty was formed in the mid-1990s, genetically modified organisms (or GMOs) were very much in the media spotlight. America's Food and Drug Administration (FDA) had just approved the slow-ripening GM Flavr Savr tomato for human consumption, and protesters were busy uprooting test fields of GM crops. But no country's government had ever approved a GM animal for human consumption. This was uncharted territory. So AquaBounty approached the FDA, who turned to Eric Hallerman, a Professor of Fish Conservation from Virginia Tech's College of Natural Resources and Environment.

I spoke to Eric, who told me about the concerns that were raised at the time. Critics were worried that the transgenic fish could be unusually allergenic, but Eric countered that because the salmon is made up of genes that come only from edible fish, the risk would be no different from eating a non-GM fish. 'If you're allergic to fish, you just wouldn't buy this product,' he says.

Then there were the fish themselves. People were anxious the salmon's unusual genetic makeup might cause it to develop health problems. 'This is a serious issue,' says Eric. 'No one wants animals to suffer.' Eric found that the fastest growing fish sometimes do have problems. 'It's rare but sometimes they can have misshapen heads and fins,' he says. Abnormalities like this crop up in normal fish stocks from time to time, but according to Eric, they are more common in the transgenic salmon. 'The answer,' he says, 'is not to use them as brood stock.' The misshapen fish never get to reproduce, so the characteristics are not perpetuated.

The biggest concern, however, was over the risk the transgenic salmon posed to the environment if they ever managed to escape. In the early 1990s, farmed salmon raised in ocean pens did sometimes slip through the net and swim off to the sea. Environmentalists worried that if the transgenic salmon did the same, they could threaten their wild relatives by competing with them for food or interbreeding with them. 'Here you have a fish that puts all of its energy into growth and none of it into reproduction or disease resistance. The worry is that if the two types of fish mix, the population might be less viable going forward,' says Eric.

AquaBounty knew from the beginning that they would be unlikely to raise their fish in ocean pens, so they set up their operations in landlocked facilities. The fertilised eggs are produced in a large, steel-framed building on Canada's Prince Edward Island and then flown to another land-based facility in eastern Panama where the fish are grown to maturity in tanks. There is no escape. On top of that, all the fish are female, so they can't breed with one another, and all the females are sterile, so they can't breed with anything else. The way this is achieved is a gloriously techy piece of science that I feel compelled to share with you so that one day if you find yourself discussing fish genomics at party – it could happen – you will totally nail it. Here goes:

In the fish world, it's a well-known fact that if you expose freshly fertilised female eggs to male sex hormones, the fish will remain genetically female but grow up to produce sperm. It's confusing I know, not least for the fish, but bear with me please. Just like humans, fish have X and Y sex chromosomes. Females have two X chromosomes, while males have one X and one Y. So the masculinised female fish – which have two X chromosomes – grow up to produce sperm that contain only X chromosomes. The sperm is then used to fertilise regular Atlantic salmon eggs, which because they come from regular females, also only contain X chromosomes. As a result all the offspring are female, but not sterile – yet. Half an hour after the eggs are fertilised, they are put in a column where they are subjected to extreme atmospheric pressure. This prevents them from discharging an extra set of chromosomes that would normally be ejected, so the resulting fish are not only female, they are also triploid: they have three sets of chromosomes instead of two. This makes them sterile. Even if they could escape from their mesh-covered, carefully filtered, landlocked tanks, they couldn't pass on their unusual genes even if they wanted to.

In the end it took US$80 million (£64 million), a lot of political wrangling and 25 years of research to prove to the FDA that the AquAdvantage salmon poses negligible risk to the environment and is safe to eat. So the question now is, will people want to eat it? In a 2013 *New York Times* poll, 75 per cent of participants expressed concern about eating genetically modified food, and although the fish is currently on sale in Canada, as of March 2019, it is still not sold in the US. Although the FDA certified the salmon safe to eat, its import has been banned until the agency determines how it should be labelled.

Pimp My DNA

The AquAdvantage salmon is an astonishing creature because it carries the DNA of not two, but three different species: a feat that is only made possible because all life on Earth evolved from the same common ancestor and as a result, all life on Earth shares the same genetic code. It doesn't matter if you are a human or a hummingbird, a begonia or a bacterium, the cellular instructions required to generate your form and keep it running are written in a universally accessible form. DNA is our common language, shared between all species on Earth. This means that the machinery inside the cells of one species can, in theory, read and interpret the instructions from any other species, regardless of what those instructions say. So the cellular machinery inside an Atlantic salmon can easily follow the genetic instructions from a Pacific Chinook salmon or an ocean pout.

In the natural world, the DNA of different species can become mixed when animals of two closely related species interbreed to produce hybrids. Pizzly bears and wholpins,* for example, crop up from time to time, but these unions are between species that share close

* A pizzly is a cross between a polar and a grizzly bear, while a wholpin is a cross between a whale and a dolphin. See Chapter 7 for more unlikely named creatures.

evolutionary origins and so are genetically similar. The tools of modern molecular biology enable scientists to mix the DNA of dissimilar organisms. Now even distantly related species can swap genetic information, and the only limiting factor is the scientist's imagination.

In the early 2000s, Zhiyuan Gong from the University of Singapore was trying to a make fish version of a canary in a coal mine. He wanted to produce a tropical fish that can detect environmental toxins and then change colour to alert people to their presence. So he fused a fluorescent jellyfish gene to a promoter that turns the gene on when metals like arsenic and mercury are present. Then he added this tailored DNA into the little silver zebrafish.

If anything, the experiment worked too well. Instead of glowing green when pollutants were added to the water, the fish proudly sported their neon raiment all the time. They were rubbish canaries, but they were lovely to look at. Singapore is one of the world's largest ornamental fish exporters, and Zhiyuan, who was trained in aquaculture research, recognised an opportunity. He patented his methods and struck up a deal with an American company, Yorktown Technologies, who then started to market the transgenic tropical fish as pets.

Now if you live in the United States, it's possible to buy brightly coloured transgenic tropical fish to live in your aquarium. Dubbed 'GloFish®', they are available in six dayglow hues, including 'Galactic Purple®', 'Cosmic Blue®' and 'Sunburst Orange®', colours that have been bequeathed to them by jellyfish and coral.

Around the same time that Zhiyuan was customising his fish, researchers from Japan and Canada were doing something similar with pigs. In the early 2000s, Japanese geneticist Akira Iritani created 'the Popeye pig' when he took a gene from spinach and added it into pigs. The gene, which converts saturated fats into an unsaturated alternative, gave the animals a better fat profile. Healthier sausages were on the cards, but consumers were not keen and the project folded. No one was able to save its bacon – healthy or otherwise.

Meanwhile, 5,000 miles (8,000km) away in Canada, researchers added a bacterial gene to pigs to create the 'Enviropig', reported to be the world's most environmentally friendly pig. The gene helped the pigs break down a normally indigestible form of phosphorous found in feed grains, meaning that their excrement contained 75 per cent less of the potentially harmful chemical. This was, reportedly, good news for the environment because when high levels of phosphorous leach into streams and rivers, it can cause toxic algal blooms that suffocate wildlife. The safety data was promising, but once again the public were not interested. Financial backers withdrew their funding and the project got the chop. All the animals were euthanised in 2012.

How is it then, that we are seemingly willing to accept GM fish but not GM pigs? It doesn't make much sense. With GM food plants now commonplace, the underlying technology is becoming increasingly accepted. Perhaps as time moves on and more GM animals receive the official regulatory stamp of approval, these seemingly arbitrary boundaries will fall. But what if scientists didn't have to mix and match the DNA of different species in order to generate the animal of their choosing? What if they could get results, not by adding foreign genes, but by altering the animal's existing genetic code?

CRISPR Critters

In the early days of modern GM, the technology was still relatively unpredictable. For every embryo that incorporated the foreign DNA, many more did not and if they did, it was hit and miss how well the imported gene worked. Although scientists knew the identity of the gene they were adding, they couldn't determine where in the recipient's genome it would land. This was a potential problem. If a new gene lands in an area that, say, controls cell division, then it has the potential to disrupt the normal pattern of instructions and lead to uncontrolled cell growth and cancer. Scientists *could* add whole genes in; they could even take whole genes out; but fine-scale alterations were elusive. What was missing was precision.

'Ever since I started working in this area, the idea of altering a gene in a precise manner to do what you wanted it to do was the Holy Grail,' says geneticist Bruce Whitelaw from Scotland's Roslin Institute. Then along came a new suite of methods that enabled researchers to modify the genome with an unprecedented level of accuracy.

Genome editors are a group of techniques that give researchers the ability to intentionally and precisely rewrite genetic instructions. As well as adding and deleting sequences of DNA, the powerful new methods mean that single DNA letters can be swapped out and replaced with a nucleotide of the scientist's choosing. CRISPR-Cas9, much talked about in the media, is the most powerful of these tools.

The system was discovered in 2012 by Emmanuelle Charpentier at Umeå University, Sweden, and Jennifer Doudna at the University of California, Berkeley. The duo were studying how bacteria respond to viral infections by making an enzyme, Cas9, that cuts up the invading viral DNA. They showed how they could direct this enzyme to cut at a specific location of their choosing, by using short fragments of genetic material known as CRISPR.* So Cas9 has been likened to a pair of molecular scissors, while CRISPR has been likened to a satnav that directs the scissors.

Although the initial work was done in a test tube, scientists quickly applied the system to living cells and the technique swept through the research community like a cat meme. 'It's been an absolute revolution,' says Bruce. 'It's a revolution because it's just so easy to work with and because the vast majority of molecular biology labs around the world have been working with it, and helping to refine the technology. It's just getting easier and easier.'

In 2016, the technique was used to restore vision in rats with a genetic form of blindness. In 2018, it was used to boost levels of a missing protein in dogs with Duchenne muscular dystrophy, and in 2019, it was used to make a monkey model of depression. It's currently

* CRISPR is short for Clustered Regularly Interspaced Short Palindromic Repeats, which is why people call it CRISPR. And it's pronounced 'crisper'.

being used to develop a method to rid the world of malaria (see Chapter 5), but its most controversial application is in the human realm, where it could be used to permanently alter the DNA of our descendants.

CRISPR has the potential, not just to treat genetic disorders, but to prevent them from being passed down the generations. The flip side is that it could also be used to 'enhance' future generations by, for example, adding in gene variants that make people smarter, stronger or more resistant to disease. This is uncharted, ethically charged territory, and it's widely agreed that the science should progress slowly, carefully and with proper regulation. Society needs to determine the sorts of changes that are acceptable, and scientists need to ensure that CRISPR is safe for human application.

So it was with some surprise, in November 2018, that the world learned that a Chinese scientist was claiming to have produced the world's first gene-edited babies. A team led by He Jiankui of the Southern University of Science and Technology of China in Shenzhen claimed to have used CRISPR to disable a key gene and make twin baby girls immune to HIV. With the spread of HIV so easily preventable, courtesy of the humble condom, it's an odd gene-editing choice to have made. Critics cautioned that the procedure could have long-term effects that are hard to predict, and six months later, a different group of scientists found that when the same mutation occurs naturally, it can shorten people's lives by nearly two years. This is playing with fire. In his efforts to give them genetic protection from HIV, Jiankui may have compromised the girls' lifespans. Human beings are not laboratory animals; this is dangerous territory.

Meanwhile, CRISPR is being used to gene-edit farm animals and crops. Researchers have already used the method to make seedless tomatoes, gluten-free wheat and mushrooms that don't brown when they are sliced. Ever since the mutation that causes double-muscling was identified back in 1997, scientists have dreamed of introducing it into other agricultural animals. Now, in the last few years, they have used CRISPR to beef up pigs, goats, rabbits and even catfish. On a farm in Uruguay, there's a whole flock of CRISPR-edited double-muscled sheep. The brainchildren of Alejo Menchaca from the Institute of Animal Reproduction in Montevideo, these are now sheep that have it all. Alejo chose to alter the Superfine Australian Merino breed, which is renowned for its high-quality wool. 'But they are absolutely not a breed for meat production,' he says. Superfine Australian Merino sheep are skinny and don't carry much meat. 'People have been trying to breed a double purpose animal that produces meat and wool for thousands of years but now CRISPR has made that possible.' The edited lambs grow more quickly than regular sheep and mature to become well-muscled animals that can be used for both their fleece and their chops.

It's a 'proof of principle' study, so these animals are not for sale, but it's easy to imagine why some people might welcome them. As Alejo points out, farmers have spent a long time trying to improve their animals via selective breeding. Now here comes a method that achieves the same goal overnight. New varieties can be made in a single generation. The double-muscled animals don't contain any new genes from foreign species; rather their existing DNA has been given the most subtle of nudges.

The mutations they carry are identical to ones that crop up naturally and are found in animals like the Texel sheep and Belgian Blue cattle. So Alejo hopes that the regulators will go easy. It took decades for the AquAdvantage salmon to receive regulatory approval, but maybe CRIPSR-edited animals will have an easier ride.

Elsewhere, scientists are using gene-editing to address animal welfare issues. Appolinaire Djikeng grew up in Cameroon where his father was a subsistence farmer. He owned a handful of domestic pigs which he sold, at strategic points through the year, to pay for his children's education. But one year, in the mid-eighties, an epidemic of African Swine fever broke out. Swine fever is a virulent disease that can cause domestic pigs to bleed to death within a week. There is no vaccine. All his father's pigs died, and the incident made Djikeng realise how easily disease can blight, not just a family's livestock, but a family's future.

Fortunately for Djikeng, his mother had some chickens she could sell, enabling Djikeng to continue his studies. Thirty years later, he's now Director for the Centre for Tropical Livestock Genetics and Health, a collaborative organisation that works with the Roslin Institute to improve livestock in the tropics. Gene-editing is one of their methods.

At the Roslin Institute, there are pens full of edited pigs. Some have been altered to resist porcine reproductive and respiratory syndrome virus (PRRSV), a devastating global disease that costs the swine industry billions of dollars each year. Others are being edited to resist Swine fever. Scientists realised that wild pigs from Africa – such as warthogs and bushpigs – have natural immunity. This is because they carry an altered version of a key immune system gene. So Bruce Whitelaw and his team have edited this African variant into domestic pigs. All they've done is change five letters of the domestic pig's 2.7 billion-letter genetic code. Now they are testing the animals to see if they are resistant to Swine fever. 'We're not trying to make bigger animals,' says Bruce. 'We're trying to make healthier animals. Our goal is to improve the welfare of farmed pigs around the world.'

In a similar vein, scientists including Djikeng are using gene-editing to make cattle that are more heat tolerant, and animals that don't grow horns. Every year around 13 million calves have their horns sawn or burned off in order to stop the animals from injuring each other and us. Naturally hornless or 'polled' versions exist for some breeds, but not all, and breeders can see the value of creating dairy and beef breeds that are genetically horn-free. With the birth of two calves, Spotigy and Buri, American biotechnology firm Recombinetics edited the polled mutation into blue-ribbon dairy cows. Now the descendants of these animals will never know the stress of having horns forcibly removed.

However, Compassion in World Farming, the farm animal welfare organisation, is opposed to gene-edited farm animals on the grounds that the methods used to create them cause suffering. 'The best way to address disease issues,' says the organisation's Director, Philip Lymbery, 'is to keep the animals in decent conditions.' Diseases spread when farm animals are kept in cramped enclosures, so the organisation views gene-editing as an over-engineered fix that will only serve to perpetuate these abuses. 'It's worrying that we haven't realised that these are just further tools of intensification,' he says.

I agree that there are problems with our food system (see Chapter 6). If these engineered animals are used to prop up an already broken system, then I question their value, but I'm open-minded to the use of gene-edited farm animals in different settings. If we could somehow ensure that big companies don't have the monopoly on these animals, and that they could be made available to smallholders, it could make a big difference to their lives. There are plenty more families like Djikeng's. African sleeping sickness, bird flu, mad cow disease ... as scientists unravel the molecular mechanisms that confer immunity, animals could be modified to resist these blights. If we leave it to natural processes – wait for a useful, randomly generated mutation to occur and then selectively breed from the animals that have it – we could be waiting for a very long time. Gene-editing could be a boon to the health of our livestock, but could it also have applications for human wellbeing?

Welcome to the Pharm

What would you do if your heart was failing and you were stuck on a transplant waiting list? If doctors could save your life by implanting a genetically modified pig heart, would you give them the green light to go ahead? In the not too distant future, this currently speculative scenario could become reality as researchers use CRISPR to modify the pig genome, and tailor-make transplant organs for use in people.

There is a global shortage of transplant organs. In the US alone, more than 110,000 people are waiting for transplants and it's estimated that every 10 minutes, a new patient joins the national transplant waiting list. For years, scientists have been mooting the possibility of putting pig organs into people. Although it may sound far-fetched, pigs are an obvious choice. They are compliant creatures that breed well in captivity and have similar-sized organs to humans, but if they were transplanted into people as they are, they would be rejected immediately.

Part of the problem is that pigs have viruses embedded in their DNA that they pass on to future generations via their sperm and eggs. Although the viruses – called porcine endogenous retroviruses (PERVs) – are not thought to be dangerous to the pigs, the worry is that, if a pig organ were transplanted into a person, the PERVs could infect the surrounding human cells and cause diseases like cancer. Getting rid of the PERVs is no mean feat, not least because there are so many of them, but in 2017, Harvard geneticist George Church and his team used CRISPR to produce 15 little gene-edited piglets, all cleansed of the viruses. Pig–human transplants moved a significant step closer.

It's an auspicious start and the first of many precise alterations that will need to be made. When surgeons transplanted a pig heart into a baboon back in the 1980s, they were shocked to find that the baboon died within a few minutes. The explanation was that pig organs are covered with carbohydrate molecules that mark the tissue for immediate destruction by baboon antibodies. So now Church is using CRISPR to make pigs whose organs lack these carbohydrates, among other alterations. The long-term vision is to combine these approaches and create the perfect pig for organ transplants.

Bruce Whitelaw from the Roslin Institute calls Church's 2017 study 'a real tour de force'. 'It's an indicator of how we will use this technology going forward,' he says. It's easy to feel a knee-jerk reaction against the idea of putting pig organs into people, but if the idea feels unnatural, just remember: regular transplantation is unnatural too – but it has still saved hundreds of thousands of lives. If the technology can be proven to be safe and beneficial, it could offer hope to those whose lives might otherwise be cut unnecessarily short. Although there's clearly some way to go before pig organs find their way into human bodies, other animals are already being used to make medicines for people.

The first GMO ever made, back in the seventies, was a bacterium that produces human insulin. The drug was quickly approved by the FDA and its production was scaled up. Bacteria can be grown cheaply in vast quantities in huge fermenters and the drug has since become a life-saving treatment for diabetes sufferers around the world.

After that, other bacterially generated medicines followed. Bacteria are good at producing large, complicated proteins that are difficult to obtain by other methods. Before GM bacteria came along, for example, scientists used to extract human growth hormone from the brains of deceased people. The hormone is used to treat people with certain forms of dwarfism, but then scientists found that recipients sometimes developed a fatal neurodegenerative disorder, Creutzfeldt-Jakob disease. Disease-causing molecules called prions were being transferred from infected dead bodies to patients via the very treatment that was meant to help them.

They needed a different way to produce the same protein. Bacteria offered a clean, hygienic way to mass-produce the same hormone and today many therapeutic molecules, including interferon for certain cancers, clotting factors for haemophilia and erythropoietin for anaemia are made in these marvellously versatile single-celled creatures.

Larger animals offer the potential to make larger quantities of medicine, so after mastering the use of bacteria, researchers were quick to realise the potential of 'pharming': genetically modifying farm animals to produce medicines. There are already a handful of drugs on the market that are made in this way. In 2009, the FDA approved an anti-coagulant drug called ATryn, made by GM goats that produce the molecule in their milk. This was followed by a drug used to treat a rare swelling disorder, harvested from the milk of GM rabbits, and an enzyme used to treat a rare, inherited condition, made from the eggs of GM chickens.

Research from the Roslin Institute highlights various benefits that come from this approach. The team there have also been 'pharming' chickens, and produced GM birds that make various medicines in the whites of their eggs. These are not ready for patients yet, but one, called IFN-alpha2a, has antiviral and anti-cancer effects, while another, called macrophage-CSF, is being developed as a therapy that stimulates damaged tissues to self-repair.

The hens, of course, have no idea how special they are. They live normal lives in large pens where they are cared for by trained technicians, who tend to their needs and collect their eggs. Lissa Herron of Roslin Technologies told the BBC, 'As far as the chicken knows, it's just laying a normal egg.' Only three eggs are needed to produce a clinically relevant dose of the drug, and bearing in mind that some hens lay up to 300 eggs a year, this means that pharming could end up being more cost-effective than other production methods.

The biggest saving comes from the fact that chickens are not demanding to look after. Chicken sheds are much cheaper to build and run than the sterile laboratories otherwise needed for factory-scale production. Once the animals are made – which is quick, easy and cheap using CRISPR – the only cost is their upkeep. Given the advantages of using animals to make complicated molecules, it's no surprise that researchers have now moved beyond the therapeutic realm.

The Bleat Goes On

Spiders are remarkable creatures. The silk that they spin has a strength and elasticity that is as yet unmatched by anything that humans have made. It's stronger than steel or Kevlar, and is so stretchy that Indo-Pacific fishermen have been known to use it for fishing. It's also insoluble in water, and doesn't generate much of an immune response when it comes into contact with human tissue. This impressive list of properties was called upon by the ancient Greeks when they used spider webs to patch wounds in bleeding soldiers, and it has led modern-day doctors to wonder if the same substance could be used in surgical procedures like ligament repairs.

There's just one problem however. How to get enough of it? Spiders are relatively small, so it takes an awful lot of them to make a meaningful amount of silk. In 2009, a rather fabulous golden spider-silk cape went on display at London's Victoria and Albert Museum. The lavish shoulder-to-floor number was delicately embroidered with spiders and wild flowers, but it took 1.2 million Golden orb-weaver spiders to help make it. They were wild spiders, collected from their natural environment in Madagascar, then released after they had donated to the project, but this approach is obviously unworkable for medicine where a regular supply of spiders would be needed. Think how much easier it would be if the arachnids could be farmed, but again, there are problems. 'Golden orb-weaver spiders have two personality defects,' says biologist Randy Lewis from Utah State University. 'They are both territorial and cannibalistic.'

So Randy came up with an alternative idea. He took a spider-silk gene and used it to create a line of transgenic goats that make spider-silk proteins in their milk. Spiders make a variety of different silks, but Randy was interested in the strongest of them all: dragline silk. These are the bungee-like threads spun by the insects to prevent themselves from falling when they create the outer frame of their webs. Dragline silk is made of short strands of protein, which line up and self-assemble as they leave the spider's spinneret. The female goats produce the same short protein strands, which float freely in their milk. Retrieving the threads is not difficult. Goat milk is naturally homogenised, so first, the cream has to be separated out using a machine and then the milk has to be filtered. After that, it's possible to dip a fine glass rod into the opaque solution, swizzle it around and lift out a single strand of silk. The thread is so robust that it can be fastened to a spool and physically wound out of the liquid.

When Randy first started manufacturing this silk, over 20 years ago, it was strong and stretchy, but not quite as good as authentic spider silk. Some goats make better spider silk than others, so over the years, Randy has selectively bred these elite animals together to produce a line of goats with superior spider silk. Freckles, the stuffed goat who greets visitors at Pittsburgh's Center for PostNatural History, is one of these, and there have now been nine generations of spider-goats. 'We have greatly improved the productive capacity of the goats, both in terms of the protein they make and in terms of the milk they make,' says Randy. Now the silk they produce can stretch up to 20 times its original length, and weight for weight, is up to 10 times stronger than steel.

This makes it attractive to surgeons looking to repair ligaments. Currently, medics repair torn ligaments by using slivers of flesh extracted from a patient's own muscles, or donor tissue that comes from cadavers. Neither solution is permanent, but spider silk could offer a tough, stretchy and enduring alternative. The applications don't stop there however. When Randy thinks spider silk, he thinks big.

Spider silk could be used to make parachutes or cables for suspension bridges. It could be used to make adhesives and gels, coatings and films. Randy has found that the silk threads can be heated to 1,800°C without burning or melting or losing their structure. 'We can basically turn them into carbon fibres,' he says. 'They're as strong as carbon fibres, but not as stiff.' So spider silk could be used to make body panels for cars or wings for drones. It could be woven into a fabric and used to create protective clothing. It won't stop a bullet, because the fabric would stretch, but it could stop the smaller particles often jettisoned by crude explosive devices. 'You could use it to make a liner for pants* or another set of underwear,' says Randy.

Not content with just spider-goats, Randy has now added the same gene into silk worms, bacteria and alfalfa. Goats are great because they make large volumes of the protein strands, but the milk then has to be treated and the fibres have to be drawn out. Silk worms – another domesticated species – cut out the middle man. They spin the spider-silk fibres for Randy directly when they make their little cocoons. 'Every generation, their silk gets stronger,' says Randy. 'The best ones we have now are as good as spiders at making silk.' Bacteria, on the other hand, are easy to scale up and grow in large vats, while alfalfa plants are readily grown and have a naturally high protein content. 'We're hedging our bets at the moment,' says Randy. 'If I knew which of these approaches was going to be the best one I wouldn't have four different things on the go.'

All of a sudden what seems like an eccentric idea is starting to sound like erudite innovation. The world needs new materials and the natural world is a constant source of inspiration. We have come a long way from the red canary to the spider-goat, via medicine-making chickens and glow-in-the-dark fish. Thanks to CRISPR, we can now rewrite the genetic code at will. If selective breeding has been used to steer evolution, then CRISPR can be used to completely derail it. The last common

---

* US pants = UK trousers.

ancestor of spiders and goats, for example, would have existed well over 400 million year ago and from that day to this the two groups of animals have never exchanged genes … until Randy Lewis came along and made Freckles. If transgenic animals lead to new materials and medicines and cures for disease, then is this really such a bad thing? We've been modifying the genomes of farm animals ever since they were first domesticated many thousands of years ago. Now CRISPR provides the opportunity to think beyond the conventional boundaries of artificial and natural selection, to generate entirely novel organisms the likes of which have never been seen in over 3 billion years of life on Earth.

CHAPTER FOUR

Game of Clones

10 December 2006

In the closing moments of the greatest polo tournament ever, the Argentine Open Championship at Palermo, the crowd were on the edge of their seats. Three minutes into the extra chukka,* the teams were level. All eyes were on Adolfo Cambiaso, the world's Number One polo player, waiting for him to whack the ball between the goal posts and secure another win for his team.

For much of the match, he had ridden one of his favourite steeds, a chestnut stallion called Aiken Cura, but the relentless pace was taking its toll. The horse was tiring so Cambiaso decided to swap mounts. He galloped down the field and led his stallion into the stabling area. Then it happened. Without warning, Aiken Cura's front left leg gave way and the horse crumpled to the ground. Cambiaso leapt from the saddle and threw his blue and white helmet to the ground in distress. 'Save this one,' he implored the assembled veterinary team. 'Save him no matter what.'

His words rang in the air as Aiken Cura was loaded into an ambulance and driven to a nearby animal hospital,

* The game of polo tends to last for a couple of hours. It is divided into shorter periods called chukkas.

where his leg was later amputated below the knee. Fitted with a prosthetic, the hope was that the elite animal could still be used as a stud, but it wasn't to be. A few months after he played so valiantly at the Argentine Open, the decision was taken to euthanise him.

Before it happened, Cambiaso had a final wish. At this point, many animal lovers ask for a few quiet moments with their companion, but Cambiaso's request was different. He had heard of Dolly the sheep and hoped that one day, scientists would be able to clone his beloved Aiken Cura. So he asked veterinarians to freeze some of Aiken Cura's cells and stow them away.

Ten Years Earlier…

It all began when Karen Walker was attending a friend's wedding in the Scottish Highlands. She returned to her hotel room to find that a fax, dated 5 July 1996, had been pushed under her door. It read as follows:

> She's been born and she has a white face and furry legs.

News of the hirsute arrival must have flummoxed the hotel staff passing on the message, but it made perfect sense to Karen. After all, she had been there at the exact moment when the new life began.

The subject of the fax message was a newborn lamb, 'conceived' – if that's what you can call it – five months earlier at an animal research centre called the Roslin Institute, not far from Edinburgh. On 8 February 1996, Karen was having a difficult day. The embryonic sheep cells she had planned to use for her experiments were not looking good, so a colleague had donated a spare flask of udder cells that came from an adult sheep. Crouched over a microscope in the cupboard-sized laboratory, Karen inserted a fine glass pipette inside one of the cells and the DNA-containing nucleus was carefully removed. Then, with an incredibly steady hand, she transferred the nucleus into a second cell: an unfertilised sheep egg that had been 'enucleated' (its nucleus had been removed). Walker then passed a small pulse of electricity across the reconfigured egg cell, which stimulated it to start dividing. One cell became two, two cells became four, four cells became eight and so it went on. A short while later the tiny embryo was transplanted into the uterus of a surrogate ewe, who carried it inside her as the new life continued to grow.

Four and a half months later, the plucky ewe was ready to pop. When the little bleater was born, at 4.30pm on 5 July 1996, the team were reassured to see she was healthy. Within half an hour she was standing and taking her first tottering steps. I've said it before and I'll say it again: it was one small step for lamb, one giant step for lambkind. This was no ordinary lamb, though it was the colour of the face that gave it away.

When the embryo was created, the udder cell that was used had come from a white-faced Finn Dorset ewe. The egg, however, had come from a Scottish black-faced sheep, and the same breed was used for the surrogate mother. The newborn's snowy visage matched that of her nuclear donor, rather than that of her surrogate or the egg donor. So the fax indicated to Karen that the newborn was a clone.

John Bracken, a research assistant who witnessed the birth, quipped to a colleague that the lamb should be named after the busty Country and Western singer Dolly Parton, because the DNA used to create her came from mammary tissue. As the world now knows, the name stuck. How could it not? It was perfect, and so much catchier than the lamb's original name, 6LL3. One yarn that has since become enshrined in folklore (if not in fact), tells how when Parton's PR team got wind of the woolly arrival's name, they announced: 'There is no such thing as baaad publicity.'

Dolly the sheep was the first mammal ever to be cloned from an adult cell. She was created by a team under the leadership of Ian Wilmut and Keith Campbell, who weren't trying to create fields full of identical sheep. Rather, in a time before CRISPR, they were looking for a way to make transgenic animals that could make medicines in their milk. The idea was that they would alter the DNA of a single cell in culture, then use that cell's nucleus for cloning. So they decided first to clone a sheep from an ordinary cell before moving on to the next step: using a cell that had been genetically modified. Dolly, whose genome was unedited, was effectively a prototype.

Looking back, it's remarkable that one small lamb could cause such a fuss. Karen's fax message was deliberately cryptic because the team wanted time to run extra tests and write up the work as an academic paper. The study was due to be published on 27 February 1997, but a few days before its official release, a journalist from the *Observer* newspaper broke the story. The news piece relayed the story of Dolly's creation and commented, 'It is the prospect of cloning people, creating armies of dictators, that will attract most attention.' They weren't wrong. What followed can only be described as unbridled media hysteria.

Reporters flocked to Scotland in their droves. Television trucks and camper vans gridlocked the Roslin car park, while indoors, the phone started ringing off the hook. In the week that followed, the researchers fielded thousands of phone calls. The newspapers made the link with human cloning and asked if Dolly was the start of a slippery slope. Who or what was next? Could Hitler or Einstein be cloned? What about pets or dead children? Campbell and Wilmut were accused of 'playing God' and some even suggested that Dolly's birth had been kept secret because the Roslin researchers were planning to clone a human. (They absolutely were not.) Claims were made that cloning would lead to new insights into human disease and the development of new therapies, and one commentator suggested that, in terms of importance, Dolly's creation ranked alongside the discovery of DNA or the building of the atom bomb.

Today, more than 20 years after Dolly's birth, human cloning remains rightfully in the realms of science fiction and the practice is currently banned in more than 50 countries. Meanwhile, Wilmut and Campbell's original vision has been achieved. Cloning *is* being used to generate medicine-making transgenic animals, as well as other laboratory animals for medical research, but I'm intrigued to explore how cloning has permeated the wider world. Thanks to the ground-breaking work of Wilmut and Campbell, artificially generated clones are no longer restricted to the lab. If you know where to look, they can be found in fields, in stables, on race courses and sometimes even in people's homes.

Aiken Cura 02

Polo is one of Argentina's most popular sports. Teams of four compete against one another on horseback. They charge around an enormous field, controlling their steed with one hand and their mallet with the other. It takes an incredible level of skill and horsemanship, not to mention cojones of steel. Adolfo Cambiaso, who grew up on an Argentinian polo ranch, has played ever since he was a youngster. When he was just 16, he became the youngest player ever to achieve a 10-goal handicap, the best there is. When he was 25, he decided to create his own polo team, La Dolfina, and start a horse-breeding business from scratch. Today he maintains his 10-goal handicap, and has won just about every polo trophy there is. He is widely considered to be the best polo player on the planet.

When Aiken Cura died, back in 2007, various animals had already been cloned. CC, the world's first cloned cat, had been born, not to mention various genetic doppelgängers of mice, rats, goats, pigs and cattle. In Italy, a researcher called Cesare Galli became the first person to create a cloned horse using DNA from an adult cell. He called the horse Prometea, named after the Greek God Prometheus who supposedly fashioned man from clay. Although he didn't fully understand the ins and outs of cloning, Cambiaso recognised the potential. He hoped that one day Aiken Cura's frozen cells could be thawed and used to make a genetic replica.

Things started moving in 2009, when he met a Texas businessman called Alan Meeker. Meeker was an amateur polo player who dreamed of cloning the world's best polo ponies in order to create his own elite team.

Cambiaso, with his world-class stable of proven winners, was the obvious person to contact and together the duo founded Crestview Genetics, a company that specialises in horse cloning. In the early days, while they set up their own laboratory, they sought the help of established experts. They approached an American company called Viagen who were already cloning horses, and sent the company some of Aiken Cura's frozen cells. A year or so later, Aiken Cura's clone was born.

Buoyed by this success, Cambiaso and Meeker started to clone other polo ponies too, including an exceptional brown mare called Cuartetera. Most polo ponies begin playing the game when they are around five years old, but Cuartetera was so skilled that Cambiaso began playing her when she was just four. She is agile, obedient, can turn on a sixpence, and has carried Cambiaso to victory many times. The team produced multiple genetic copies. When one of them was put up for sale at a Buenos Aires auction, the three-month-old foal fetched a whopping US$800,000 (£640,000). Cuartetera's clone became the most expensive polo pony in history.

At this point, it's worth noting that no one knew how well or otherwise cloned horses would perform in a game. The animals that had been produced so far were young and untested. People were intrigued, but critics wondered whether a clone could compete at the same elite level as the original animal.

Then in 2013, Cambiaso saddled up a different clone in preparation for the final of the Palermo Open. The mare was Show Me, a copy of a successful American thoroughbred called Sage. The chestnut mare performed beautifully, just as Sage had done before her, and

La Dolfina claimed another victory. It was, perhaps, a defining moment for the history of polo. Cambiaso proved, without a shadow of a doubt, that cloned horses have the capacity for greatness too. Show Me showed them.

More was to come. Polo is a fast-paced game that asks a lot of the horses that are used to compete. Players are allowed to change their mounts frequently. Sometimes a competitor will ride eight or more different horses in just one game. In the final of the 2016 Argentine Open Championship, Cambiaso upped the ante when he rode not one but six clones. Named Cuartetera 01 through to 06, the horses were all copies of his beloved mare of the same name. They performed magnificently. Media reports suggested it was as if the clones were born to play polo. One commentator told Argentine TV, 'for a polo player to ride an extremely talented horse, like Show me or Cuartetera, would be like for a soccer player to wear Maradona's feet.' Cambiaso's team took home the title for the sixth time.

It was a shrewd and clever move. If I were being cynical, I would call it nothing short of PR genius. Cambiaso is the world's most high-profile polo player and the Argentine Open Championship at Palermo is the world's most famous polo match. The competition provided Cambiaso with the perfect platform to showcase, not just his skill, but the product that his company manufactures: elite polo pony clones. Argentina's ultra-rich polo-playing elite were impressed and Cambiaso's involvement was the ultimate endorsement. If the world's best polo player was using clones in competitions, then surely, this was the way to go.

Since then, an undisclosed number of wealthy individuals have had their polo ponies cloned. Crestview Genetics now operates from a custom-built lab, about an hour outside of Buenos Aires, where a team of dedicated researchers perform cloning experiments three times a week. To date, the lab has produced more than 70 healthy cloned horses, including at least 14 copies of Cuartetera, and Crestview Genetics now has competition from rival Argentine cloning company, Kheiron Biotech.

Outside Argentina, horse cloning is now commercially available in Brazil, Colombia, Italy, Australia and the USA, and the practice is not restricted to polo ponies. Thoroughbreds have been cloned, as have jumping horses. In the US, where rodeo-style competitions attract big crowds, researchers have cloned bucking horses and barrel-racing horses.* Quarter Horses have been copied. In Brazil, scientists have cloned two native breeds, the Campolina and the Mangalarga Marchador, while elsewhere scientists have created genetic replicas of Haflingers, Arabians and Warmbloods.† Today it's thought there are over 375 cloned horses in existence.

Although they are still very much in a minority, cloned horses are becoming increasingly common on the polo field, where they continue to divide opinion. Critics dislike their perceived artificiality, and argue that the animals offer their owners an unfair advantage that

---

* Barrel racing is a rodeo event in which horse and rider race to complete an obstacle course of barrels laid out in a cloverleaf pattern.

† Warmbloods are a group of middle-weight horses. They are often used in horse trials.

can persist well beyond the clone's original lifetime – clones of clones, and clones of clones of clones, could keep the winning streak going indefinitely. Yet the body that governs Argentinian polo is remarkably relaxed. The Argentine Polo Pony Breeder's Association (AACCP) states that its goal is 'to promote the research and the application of artificial insemination, embryo transfer and any other technique to improve animal breeding'. Basically, it permits any breeding technology that will elevate the level of play, cloning included.

Elsewhere, the organisations that govern other equine competitions have their own rulings. The Professional Rodeo Cowboys Association allows clones to compete in barrel-racing and donkey-racing events, and in 2012, the International Equestrian Federation, the governing body for equestrian sports, lifted a ban prohibiting the use of cloned horses in the Olympic Games. There is now no reason why cloned horses cannot compete against their traditionally bred counterparts in the famous competition.

Breed registries, by and large, are more conservative. In order to be officially recognised as a member of a particular breed, horses need first to be registered with the appropriate breed registry. In the US, for example, Quarter Horses are registered by the American Quarter Horse Association (AQHA), while Arabian horses are registered by the Arabian Horse Association. To become accepted, the mother and father of a prospective member must both be on the register. In the US and elsewhere, horses are only valuable if they are registered, but clones throw a spanner in the works. Clones don't have a mother or a father; not in the traditional sense. They have a surrogate and a nuclear donor. They don't fit neatly within any traditional family tree, so breed registries are not keen to accept them.

This has not gone down well with the people who have paid a small fortune to have a horse cloned. In 2012, Texas horse breeder Jason Abraham and veterinarian Gregg Veneklasen took legal action against the AQHA, claiming that a cloned Quarter Horse called Lynx Melody Too should be allowed to join the register. They suggested that the organisation's ban on clones violated federal antitrust laws, and won the case when a Texas jury decided in their favour. Clones were in! Then the case went to a Court of Appeal, and the decision was overturned. Clones were out!

Today most breed registries* refuse to admit clones or their offspring. This leaves many clones in a state of unofficial limbo, and many wealthy owners in a state of teeth-grinding annoyance. Elsewhere, the 2012 and 2016 Olympic Games came and went without a clone in sight. The competitive equine world has not been inundated with clones. The genetic replicas are still very much in the minority and traditionally bred horses still make up the vast majority of sporting animals. So why are some people prepared to spend upwards of £80,000 (US$100,000) to have their horse cloned?

Aiken Cura's Legacy

Most people who clone horses don't do it to produce a competition winner. They do it to replicate a valuable

---

* The breed registry for Lipizzan horses, the likes of which perform at the famous Spanish Riding School in Vienna, Austria, do admit clones under a specially added Appendix.

breeding animal. Today, you won't find Aiken Cura's clone on the polo field. Instead, he can be seen galloping through lush green fields at a place called Los Pingos del Taita in Córdoba, southern Spain. Los Pingos del Taita calls itself a 'spa for the stars' – the stars being of the equine variety. According to their website, Aiken Cura E01 'stands in service'. This is a horsey term that means the animal is a stud. The staff at Los Pingos del Taita collect, store and export the semen and embryos of valuable horses.

Cloning has become a way to preserve the unique genetic makeup of elite animals, be they polo ponies, racing horses, bucking broncos or members of an entirely different species. If an animal has died, then cells taken during its life (or shortly after its death) can be used to create clones. These clones are, more or less, identical twins to the original animal, so when the clones grow up, the sperm and eggs they produce will also be more or less identical. By creating genetic copies, cloning offers a way to extend the reproductive potential of an original animal well beyond its natural lifespan. This is one use of cloning.

Another is to help castrated animals to have offspring. Stallions and bulls are notoriously high-spirited, so castration provides a means to control their wayward hormones and behaviour. In the equine world, most competitive stallions are geldings, meaning they have been castrated and so cannot sire foals. If these animals are cloned, however, then their non-castrated copies can have offspring on their behalf. This is important. In the equine world, the males that are left intact are those that already have an established pedigree. They are the animals descended from breeding stock known to have desirable genes. By allowing these lines to reproduce, their genes become well represented and broadly distributed. Geldings, in contrast, are genetic dead ends. From a breeder's perspective, it's all well and good having a Grand National winner, but if that horse is unable to breed, then its winning genes will die with it. Cloning offers a way to perpetuate the genes of these valuable geldings. It provides a way for breeders to incorporate desirable genes that would otherwise be excluded from the breeding population. Here, cloning actually broadens the gene pool.

In the world of polo, where both male and female horses are used, cloning has an additional use. Sometimes players are reluctant to use their high-performing females for breeding. They see it as a distraction from the game. If the elite mare is cloned, however, then play can continue uninterrupted while her clones see to the business of producing foals.

Cloning has become a tool of selective breeding. Although it's not intuitive to see how making a genetic copy can alter the genetic makeup of individuals over time – which is what selective breeding does – when you reduce the practice to its basic function the goal becomes apparent. In this instance, cloned animals are simply vessels that allow large quantities of sperm, eggs and embryos to be created. From these cells, new animals are made. So cloning serves to perpetuate key characteristics that are deemed valuable, such as trainability, agility and stamina, by enabling clones with these features to have more offspring.

Aside from their cloning work, at Crestview Genetics, clones of elite mares are mated to sires of high genetic value, in the hope that their offspring will become future polo champions. Over 200 foals have been created in this way. It's about mixing together the best possible DNA that there is, from past winners with proven track records. Cloning does not make an animal immortal because it will still die, but for as long as its cells exist, its undiluted genome has the potential to continue. Stored correctly, with protective anti-freeze chemicals, frozen cells can remain viable indefinitely. As long as the cells are available, there's nothing to stop a researcher in 2050 cloning a horse that is alive today. Cloning provides a means to link individuals that are separated by time.

Bully For You

In the cattle world, cloning is similarly seen as a way to preserve the genetics of elite animals. Cattle were first cloned way back in 1997, just a year after Dolly was born. The first clone, called Gene, was born at the American Breeders Service facilities in DeForest, Wisconsin, USA. Since then, thousands of cloned cattle have been produced, most of them high-value animals that breeders just can't afford to lose. Some bulls, for example, are worth hundreds of thousands of dollars, so cloning, which costs a few tens of thousands of dollars, provides a means of insurance. When the original dies, the clone can live on.

One famous bull, a jet-black Angus called Final Answer, was 2,475kg (5,500lb) of pure muscle. Unlike the majority of beef cattle, who are destined to be doused in gravy, Final Answer's job was to produce semen. Lots of it. With a scrotal circumference of 44cm (17.3in, roughly equivalent to the pumped biceps of a

20-year-old Arnold Schwarzenegger) Final Answer produced over 500,000 'units' or shots of semen during his life. That's more than 100 units per day. Exhausting! His seed was sold and used for artificial insemination, and such was his reputation that Final Answer became one of the most prolific sires in the beef cattle industry. This one bull fathered hundreds of thousands of offspring. Faced with the prospect of Final Answer's finality, breeders decided to clone him. So when Final Answer did finally die in 2014, Final Answer II took over. In recognition of the fact he is not the original, today, a sample of Final Answer II's sperm retails for around US$22 (£17.50). That's half price compared to a shot from the real McCoy. It's a total bargain if you're looking to buy that sort of thing. Final Answer II offers the exact same genetic material, but at a fraction of the cost.

With such esteemed ancestry, Final Answer II's immediate future is certain. He will continue to pump out semen until he too expires. After that, who knows? Maybe there will be a Final Answer III or IV or V. In China, however, the fate of cloned cattle is very different indeed.

In 2016, a Chinese company called Boyalife announced that it planned to build an enormous cloning factory in the coastal city of Tianjin. When production is in full swing, researchers expect to churn out a million cloned cattle every year, but their animals will not be used as sires. Instead they will help to supply China's rocketing demand for beef. If Boyalife have their way, there will indeed be fields full of cloned cattle, and meat from them will be ending up on the dinner table.

When I mention this to friends, it prompts a common reaction: 'Yuck!' In the UK at least, people feel uncomfortable with the idea of consuming products derived from cloned animals. They say that it feels unnatural, but in fact, the meat and milk is no different from regular animals. These animals are not genetically modified. Their DNA has not been altered; rather, it has been copied. They are twins and triplets to the original animals.

In recognition of this fact, the US Food and Drug Administration (FDA) has ruled that meat and milk from cloned animals and their offspring are as safe to consume as food from conventionally bred animals. It is legal to sell meat and dairy from cloned livestock in the US. In the UK, clone-derived foodstuffs are also legal, although they are classed as 'novel foods' so special permission is needed before they can be sold. The reality, however, is that you're unlikely to have sipped the milk or chomped the rump of a cloned cow. Compared with the cost of producing a regular dairy or beef herd, cloning is not an economical way to produce milk or meat, so instead the clones are used for breeding. That said, if you've ever been in the US and consumed milk or cheese, it's highly likely that the dairy products came from the descendant of a cloned cow.

If Boyalife have their way, China will be the first country to mass-produce cloned cattle. By scaling up production to industrial levels, the cost should come down and the company envisage supplying 5 per cent of the country's premium beef cattle. 'In China we do things on a massive scale,' Xu Xiaochun, Chief Executive of Boyalife, said in a 2014 interview. 'We want to do all this not just for profit, but also for history.' Cloning is becoming big business. Given the potential profit it is perhaps no surprise that some of the larger cloning companies have now added another string to their bow. If you have a lot of money to part with, they will offer to clone your pet.

Higgs 02?

In March 2018, the *Guardian* newspaper reported that Barbra Streisand had had her pet dog cloned. Two copies were, reportedly, produced after her 14-year-old Coton de Tulear dog, Samantha, had died the year before. For those not familiar with the Coton de Tulear breed, they are adorable, small white balls of fluff. The actress/singer named her new dogs Miss Scarlett and Miss Violet, and dressed them in red and lavender so that she could tell them apart.

Two years earlier, Belgian fashion designer Diane von Fürstenberg paid a small fortune to have her Jack Russell terrier, Shannon, cloned. Two clones were made, which she named Evita and Deena. Just like Shannon, both pups have a dress named after them. The Evita is a 'universally flattering' sleeveless number designed to sculpt the hourglass figure, while the Deena is a loose-fitting silk frock in 'sea garden' with a V back and ruched cuffs.

I decided to approach Viagen, the company that produced Streisand's cloned dogs, to find out how I would go about cloning my dog, Higgs. To date the company have produced more than 100 healthy cloned dogs. While I am fascinated by the ability of scientists to clone cattle and horses, the worlds of agriculture and equine sport are remote to me, but as one of the world's 32 million dog owners, dog cloning has an emotional dimension for me that colours my perception. In this case, scientists are not replicating an animal for breeding purposes or for sport, but to replace a lost companion. It makes me feel very uneasy.

I filled in the 'Request for Free Information' form on the company's website, then found myself dithering. Higgs was snoozing at my feet. He must have been chasing a squirrel in his dreams, because his feet were twitching and he was 'yipping' softly. I reached down and tickled him behind the ears. The trembling stopped and the noise subsided to a gentle snore. Gazing down at him, I found it hard to imagine there could ever be another Higgs. More than that, I felt guilty for even thinking about it. How could an animal so unique, so special and so loved *ever* be replaced? To contemplate ordering a copy felt like betrayal. When I eventually pressed the 'submit' button I felt like a total scumbag ... so I gave him a piece of cheese.

Dogs were first cloned back in 2005, when South Korean cell biologist Woo Suk Hwang produced a black and tan Afghan hound called Snuppy (short for Seoul National University puppy, where Hwang worked at the time). Hwang now works in a shiny new lab that he founded on the outskirts of Seoul, called the Sooam Biotech Research Foundation, where they specialise in cloning large animals, such as dogs, cattle and pigs. Since the lab was established, Hwang and colleagues have churned out more than 500 cloned dogs. Some of these are family pets. Others are copies of highly trained police dogs. With many dozens of these animals in service, if you ever pass through the baggage reclaim at Seoul Incheon Airport, the chances are that your luggage will have been inspected by a cloned sniffer dog.

Twenty-four hours after I sent my enquiry to Viagen, I received a reply. It was a breezy email from one of Viagen's Client Service Managers. She thanked me for my interest in ViaGen Pets and explained how the first step would be 'Genetic Preservation'. For US$1,600 (£1,275) (plus US$300, £240, for international shipping) ViaGen will FedEx me a biopsy kit so a local vet can collect a few small skin-punch biopsies from Higgs. When this is done, the sample is returned to their Texas lab where the cells are grown in culture, then frozen. When I am ready to move forward with cloning, they will use the cryo-preserved cells to create cloned embryos and within months I will be able to welcome a puppy that is a genetic twin to Higgs. If I were to proceed, the whole experience would set me back US$50,000 (£40,000),* but could it ever be worth it?

The True Cost of Cloning

Cloning is a notoriously inefficient process. When Dolly the sheep was born, over two decades ago, she was the one success story in a litany of failed attempts. Researchers created 277 cloned sheep cells. Of these, only 29 began to develop normally and were implanted into surrogate ewes, and only one healthy lamb was born – Dolly. Since then, as methods have been adjusted, the odds have improved, but not by much, and the efficiency of cloning

---

* If I had a cat to clone, it would set me back half this amount. Apparently this is not because cats are only half as good as dogs, but because the canine reproductive cycle is more cryptic than its feline counterpart. The extra money goes towards funding research to improve the cloning process.

varies considerably between both laboratories and species.

A study by researchers from Seoul National University in Korea scoured the academic literature for reports of dog cloning. They found twelve studies, published over seven years, that all used slightly different methods. In these studies, the percentage of clones implanted into surrogates that were still alive at birth ranged from less than 1 per cent to just 4 per cent. The take-home message is clear. Most cloned dogs die before they are born.

I spoke off the record to a representative from a different cloning company, who described some of the health problems suffered by cloned dogs. 'Sometimes they are born with a cleft palate, with thickened necks and with large tongues,' he told me. Sometimes the clones of male German Shepherd dogs are born with female genitalia. Their reproductive organs are deformed and the animals are infertile. No one knows why.

Problems exist in other species too. Cloned horses, for example, are sometimes born with crooked legs and can be very weak at birth. Sometimes the umbilical cord is larger than normal, which leaves the foal more susceptible to infection. The range of health problems experienced by some cloned animals range from the minor and the treatable, to the major and the incurable.

All in all, it can sometimes take hundreds of cloned embryos to produce just one or two healthy animals, and let's be clear what we're talking about here. There is death at every stage of the process. Cloned embryos die while they are microscopic cellular bundles floating around in a petri dish. Cloned embryos die while they are developing in the surrogate's uterus, and cloned animals die or have to be euthanised, sometimes, after they are born. Life isn't peachy for the surrogate either. Compared to non-clone pregnancies, surrogates have a higher rate of various health problems.

That said, there are researchers out there, working in academia and in private companies, who are working hard to improve the odds. It's thought the problems occur when the DNA used for cloning isn't fully reprogrammed. The clone inherits, not a blank slate, but a mostly blank slate with a few instructions left over from the adult cell. These confuse and disrupt the developmental process, creating health problems and spontaneous abortions.

Biologist Katrin Hinrichs from Texas A&M University was among the first to clone a horse, a little over a decade ago. Since then, she has been trying to improve the efficiency of the method. 'I'm a horse person,' she tells me. 'I don't want to be responsible for producing a foal with health problems. I don't want to foal out 10 mares and get two healthy foals. I want to foal out two mares and get two foals because I don't want to be responsible for 8 sick foals that have to be euthanised.' To achieve this, Katrin scours the academic literature to look for anything, in any species, that improves the success rate of cloning. She's been told people think that she mollycoddles her cloned embryos because of the attention she pays them, but her results are encouraging. For every three to four embryos that are implanted into three to four surrogates, about two mares will get pregnant and one will have a live foal. This is encouraging. Katrin's methods hugely increase the odds of producing a healthy cloned foal. In the right, responsible hands – with time, patience and expertise – cloning doesn't have to be such a wasteful and inefficient process.

Behind the closed doors of private companies, it's hard to know how many cloned animals fall by the wayside in the quest to produce that one perfect clone for a paying customer. Private cloning companies, by and large, do not contribute to the openly available, peer-reviewed collection of academic literature. They keep their proprietary technology close to their chests. This is frustrating. Crestview Genetics tell me that 90 per cent of established pregnancies result in a healthy foal. Viagen tell me the success rate for dogs is high, and that it only takes one round of nuclear transfer experiments to achieve a positive pregnancy and the birth of a healthy cloned puppy. All of the private companies exalt their high-tech labs and flaunt their success stories, but very few reveal the nitty-gritty of what it is that they actually do.

If I were to clone Higgs, I would want to know exactly how many copies of him it would take in order for his twin to be born. Anyone considering having their pet cloned should ask themselves first, how they would feel about having one or more unhealthy replicas of Higgs, or Milly, or Monty, euthanised in order to get the one animal they so desperately desire. I don't want that sort of suffering on my conscience, not when it is entirely avoidable and not when there are some 3.5 million shelter dogs all desperately in need of a home.

Identically Different

Then there's the matter of how similar or otherwise the clone will be to the original. Contrary to popular opinion, a clone is not genetically identical to the animal it was created from. There are small but important differences. During the process of nuclear transfer, the

DNA inside the nucleus is transported from one cell to another, but the nucleus is not the only source of DNA. Tiny energy-generating structures called mitochondria also have their own DNA. It may only be a tiny amount – around 0.0005 per cent of the total DNA content in the cell – but it's still significant and is passed, undiluted, down the maternal line from mother to daughter via the mother's egg.

This means that clones have the nuclear DNA of their DNA donor, but the mitochondrial DNA of their egg donor. This disparity means that right from the beginning clones are not exact genetic copies. Then, as time goes on, their DNA becomes more different. Living things accrue mistakes to their genetic code as their cells divide and errors are made copying the genomic blueprint. Although cells contain proofreading mechanisms to correct these errors, sometimes mistakes slip through. Certain aspects of the environment, such as pollution and radiation, can also cause changes to DNA, and although overall the genomes of clone and original will still be incredibly similar, sometimes little differences can have large effects. An unfortunately placed mutation in a gene regulating cell division, for example, could see a normal cell start to proliferate wildly, leading to the beginnings of a cancer. This can explain how sometimes, identical twins develop different genetic diseases. Similarly, animals and their clones can develop different problems.

Living things are the product of their DNA, the world they live in, and the complex interaction that is ongoing between the two. Certain characteristics, such as size, shape and colouring, are strongly influenced by genetics. You will, no doubt, have committed to memory the scrotal circumference of the prolific beef sire, Final Answer. A trip around his testicles, you will remember, measures 44cm (17.3in). The scrotal circumference of his clone, Final Answer II, is 44.8 cm (17.6in). Both animals look more or less the same, weigh more or less the same, and are similarly proportioned.

Sometimes however, nature throws a spanner in the works. When the world's first cloned cat, CC, was born, people were more than a little amazed when the original and clone bore only a passing resemblance to one another. CC was white with a grey tabby pattern, but the animal she was cloned from, Rainbow, was a calico cat with black, white and orange fur. Inside the nuclei of their cells, both animals contained identical DNA, so it seemed as if the felines were being deliberately difficult … but then that's cats for you.

So why the superficial difference? The answer lies in epigenetics, a suite of processes that alter the activity of genes without changing the sequence of the genetic code itself. Calico cats are always female, so they contain two X chromosomes. In Rainbow, one X chromosome would have carried the gene for orange coat colour, while the other would have carried a different version of the same gene that results in black fur. In the cells of female cats, one of the X chromosomes is randomly switched off. It's an evolutionary strategy to ensure that female cats – and indeed all XX female animals – don't receive a potentially dangerous double dose of any X chromosome genes. In Rainbow, the orange variant triumphed, while in CC, the orange variant was silenced, so although both cats contained the same two X chromosomes, their overall colouring was different.

Take a look at Diane von Fürstenberg's cloned Jack Russells, Evita and Deena. They look different. They have different patterns of markings and I think this is why they inspired two such very different frocks. Adolfo Cambiaso now has more than 14 clones of his champion steed, Cuartetera, and although they all look superficially similar, when you inspect them more closely, you'll notice that the white markings on their faces vary in shape and position. Some sport the odd white sock, while others have brown legs. Once again, although the clones carry the same nuclear DNA as the original animal, patterns of gene expression have changed. Subtle differences experienced by the young animals as they slumbered in their surrogate's womb altered the activity of key genes, which in turn altered the way their pigment-containing cells developed. If I were to have Higgs cloned, his genetic twin would be unlikely to sport the same pattern of black and white markings that make the original dog so dapper.

Nor would his personality be the same. Although personality and behaviour have a genetic component, they are largely influenced by environmental factors. The way an animal is brought up and the experiences that it encounters all contribute to its disposition. In some instances, if the environment is very tightly controlled, then aspects of the cloned animal's personality and behaviour might be similar to the original. When the South Korean police force train up their cloned dogs, they expose them to the same diet, physical environment and training programme as the standard police pooches in order to minimise these environmental variables. Similarly, Cambiaso goes to great lengths to raise his clones like the originals. The trainer is kept the same. The trainer's outfit is kept the same. If the original horse liked to hang out with a particular dog, then the same dog gets to hang out with the replica.

Most of us would never have the time or inclination to pander to this level of detail, which is why most cloned animals will never behave exactly like the originals. Animals can be cloned, but their personalities cannot. We are all far more than the sum total of our DNA. This is why human clones – or identical twins as they are more commonly known – grow up to become different people with their own distinct, idiosyncratic personalities. When Boyalife fill their fields with herds of genetically identical cattle, they will still form a hierarchy just like any normal field of cows. Some cows will be more bossy, others more subservient. The full spectrum of behavioural traits will be present, from timid and shy to curious and inquisitive. Each animal will have its own unique, individual personality.

It's for all of these reasons that I would never have my dog cloned. I accept that there will simply never be another dog like him, and it reminds me that I should enjoy his company now and try to find time to slip in an extra walk.

Celia, You're Breaking My Heart

It would be impossible for me to finish this chapter without mentioning one final use of cloning. Several research groups are using cloning to try to bring back extinct species. They call it de-extinction and I have written an entire book on the subject,* so I'll be brief. If

* It's called *Bring Back the King: The New Science of De-extinction* (Bloomsbury Sigma, 2016).

a species is not long gone, if some of its cells still exist, and if it has an evolutionarily close living relative then in theory it might just be possible to de-extinct it.

In 2003, Spanish scientists turned centuries of dogma on its head, when they briefly brought back an extinct variety of mountain goat called the bucardo. Cells taken from the last living individual, an elderly female called Celia, were collected and frozen while the animal was still alive. A couple of hundred embryos were implanted into 50 or so surrogate goats, of which only seven became pregnant. From these pregnancies, just one cloned bucardo was born, but although the little kid looked healthy, she died from respiratory problems just a few short minutes after she was delivered. The bucardo is not just the first animal to be de-extincted. It is also the first animal ever to go extinct twice.

Undeterred, two different groups are attempting to use cloning to de-extinct that most iconic of Ice Age animals, the woolly mammoth, while an Australian team are using the same method to resurrect a bizarre species of amphibian called the gastric brooding frog. As its name suggests, the little frog brooded its young in its stomach. Females swallowed fertilised eggs, then let the tadpoles develop internally before burping up fully formed froglets.

In some cases, cloning technology is being supplanted with CRISPR. A third group of would-be mammoth makers are recreating the behemoth – or something that approximates it – by CRISPR-editing mammoth genes into elephant cells. The aim is to produce a cold-adapted elephant with long, insulating fur and thick rolls of body fat. It will have mammoth haemoglobin, which works well at low temperatures, and little ears to minimise heat loss and the risk of frostbite. In a similar vein, CRISPR is also being used to bring back the passenger pigeon, a rosy-breasted bullet of a bird that was once the most abundant bird in North America. Ben Novak, currently at Australia's Monash University, plans to edit passenger pigeon genes into the bird's closest living relative, the band-tailed pigeon, and then use some nifty cell biology to recreate the bird.

There is much basic science still to be done but if the scientists pull it off, they will have achieved something remarkable … but it won't be the evolutionary U-turn that the media like to make out. This is something more subtle and nuanced. Scientists won't literally be bringing animals back from the dead, or recreating carbon copies of extinct life forms. Instead, they are making subtle changes to the genomes of living creatures to make them more like their extinct ancestors. The new creatures will be similar but not identical to the originals. Novak's bird will not be *the* passenger pigeon. It will be a hybrid modern-day facsimile. In the same way, if attempts to edit mammoth genes into Asian elephant cells bear fruit, then the animal that may one day emerge will not be a woolly mammoth. It will be a transgenic elephant with added mammoth genes. The creature may look like a mammoth and shrug off the snow, but it can never be the same as those shaggy beasts that saw out the last Ice Age. De-extinction doesn't mark a return to bygone eras; instead it marks an entirely new phase in the evolutionary story.

It may seem pointless to de-extinct creatures if they can never be the same as the originals, but I believe there are many good reasons why de-extinction is worthy of our attention. Humans are largely responsible for the ongoing demise of the natural world, so maybe we have a moral obligation to try to reverse the damage we have done. De-extinction could help move us towards this goal.

Along the way as the technology is developed, fundamental insights into cell biology, embryonic development, health, disease and ecology will be gleaned. Gastric brooding frogs, for example, must have stopped secreting stomach acid to keep from digesting their offspring. If we can learn how they did this, it could lead to new treatments for stomach ulcers. We can't predict exactly what we will learn in the quest for de-extinction, but the knowledge we accrue will not be wasted. It will ripple through related disciplines and bring insight to different fields.

The greatest argument in favour of de-extinction, however, is the impact the technology could have on the world's ecology. The passenger pigeon was a unique animal. It was unique in its genetics, its appearance, its behaviour and its ecology. A nomad by nature, it roamed the deciduous forests of eastern North America, gorging on the acorns and the beechnuts that it found. When the preposterously large flocks came to roost, they broke branches, felled trees and carpeted the exposed forest floor in a deep layer of guano. The scene was apocalyptic, but from this apparent devastation, life emerged. The guano fertilised the ground, prompting grasses and flowers to grow, which attracted insects and reptiles and later birds, small mammals, grazers and predators. The closed canopy forest turned into an open sunlit nursery. Passenger pigeons helped drive the regeneration of North America's deciduous forests. No other living bird can do this. If we bring back the passenger pigeon – or an updated version of it – it could boost the health of resident ecosystems. Similarly, if mammoths are ever introduced to the wastelands of Siberia, they too could help make their homeland more biodiverse (see Chapter 11).

Some find de-extinction unnatural. The scientists responsible have been accused of 'playing God', but don't we also play God when we destroy forests, over-hunt, pollute the planet and warm our world? Aren't these events unnatural too? It would be unnatural if CRISPR was being used to make horned horses (aka unicorns) or winged lizards (aka dragons). It would be unethical too, but no one is seriously considering these flights of unprincipled fancy. De-extinction is not about creating freaks of nature or icons of mythology; it's about creating healthy, genetically vibrant populations of animals that could live in the wild and contribute meaningfully to the environment. If de-extinction can in future be used as a force for good, as a tool of conservation to boost the health of ailing ecosystems, then surely this is worth exploring.

Dolly's Legacy

Let's just recap for a moment. Dolly the sheep was born over 20 years ago, and since then many other species have been cloned. Reproductive cloning – where a genetic copy of an animal is made – is being used to preserve the unique genomes of elite animals and to save these animals from reproductive toil. Cloning is being used to make high-end beef cattle, working animals like sniffer dogs and competitive animals like polo ponies. If the private companies are to be believed, cloning offers a way to lessen the grief experienced when your beloved pet dies, and the same method is also being used to bring back extinct species.

I don't want to leave you with the impression that cloning is routine or commonplace. In the wider world of animal breeding, it's a niche method practised only by those with the funds to afford it. Although cloned animals exist, they remain very much in the minority and their most important legacy is not to be found in any of the areas mentioned so far.

Imagine a scenario where someone has a devastating medical condition that cannot be cured, such as Alzheimer's disease or chronic heart failure. Now suppose researchers take some of the patient's skin cells, remove the DNA and use it for cloning. When the cloned embryo is still a tiny, amorphous mass, individual cells can be harvested and multiplied in culture. When the right mix of nutrients is added to the dish, these so-called stem cells can be coaxed to divide into more specialised cells such as brain cells or heart muscle cells. Perfectly matched to the patient they come from, these cells could then be used as a source of tissue for repair.

During the nineties, when I worked in stem cell research, scientists were hopeful that 'therapeutic cloning', as it is called, would lead to new treatments for incurable diseases, but critics were opposed to the idea because it meant that human embryos would be destroyed along the way. Then, in 2006, a man called Shinya Yamanaka from Japan's Kyoto University developed an alternative way to make stem cells. He took skin cells from an adult mouse, added a handful of genes, and reprogrammed the resident DNA into a more youthful, stem cell-like state. The cells that were produced, known as induced pluripotent stem (iPS) cells, could then be coaxed to become other cell types, such as neurons and heart cells. No embryos were harmed along the way, and for the first time, researchers had an ethical way to make replacement cells for therapeutic purposes.

The discovery earned Yamanaka a Nobel Prize, and today iPS-derived cells are moving closer to human clinical trials. In 2017, Japanese researchers demonstrated that iPS cells can relieve symptoms in a monkey model of Parkinson's disease.

This is Dolly's legacy. Her arrival opened the door to a world where cloned animals and stem cells are being used to improve our understanding of human disease, and to generate new therapies for devastating disorders. Meanwhile Dolly is on display at the National Museum of Scotland, where her stuffed remains are an inspiration to adults and children alike. Dolly is still a superstar. When I talk to schoolchildren, everyone knows her name. They know why she is special. Dolly makes people smile. There is, in her, something soft and appealing that makes people want to reach out and touch her. Literally. In recent years, she has had to be placed inside a glass case to stop people from stroking her and pinching bits of her wool. Dolly the Sheep – Rest in Fleece.

CHAPTER FIVE

Screwworms and Suicide Possums

In the late 1950s, a small number of single-engine aeroplanes took to the skies over the south-eastern United States. They flew low over the cattle-growing regions of Florida, Georgia and Alabama, and when their target was spotted, the payload was unleashed. Millions of genetically altered insects were dropped into fields containing thousands of cattle. The insects, in pupae form, duly hatched and began mating with the local flies. It was the start of one of the most successful pest eradication campaigns the world has ever known.

The campaign was aimed at screwworms: gruesome creatures that feed on the flesh of live warm-blooded animals like cows.* Adult screwworm flies are attracted to open wounds. Scars caused by dehorning, ear-tagging and castration are all fair game, as are natural scratches and the unprotected navels of newborns. Female flies lay

---

* As its Latin name, *Hominivorax* or 'man-eater' suggests, screwworms will also eat human flesh. A 12-year-old girl who was infested by screwworms while travelling in Colombia in 2008 had 142 larvae extracted from her scalp. Elsewhere, doctors have laid bacon fat over infested wounds to draw the embedded screwworms up to the surface of the skin. They call it 'bacon therapy'.

hundreds of eggs in the exposed flesh, then when the maggots hatch, they burrow their way into the injury and feast on the surrounding tissue. With their ridged bodies, they look like tiny screws. The flesh liquefies and if the wounds are left untreated, the host can die within weeks.

In the early twentieth century, screwworms were creating havoc in the beef industry, where they were chomping through more than US$100 million-worth (£80 million) of livestock a year. A scientist called Edward F. Knipling came up with a revolutionary idea to solve the problem. He decided to try contraception for flies. Knipling realised that if he could sterilise vast quantities of male screwworms in the lab and then release them into the wild, the males would seek out the regular fertile females and mate with them. Eggs fertilised by their sperm would never hatch, so if enough males could be deployed, the population would crash.

His contemporaries baulked at the temerity of the idea, but Knipling was undeterred. He grew up on a farm in Texas and had witnessed the devastation caused by screwworms. Working with his colleague Raymond Bushland, he devised a way to rear millions of screwworms in enormous dustbins full of warm minced beef, along with a method to sterilise the insects using radioactivity. Male screwworm pupae were loaded into metal tubes, then exposed to Cobalt-60. It was the insect equivalent of Atomic Gardening (see Chapter 3). Gamma rays emitted from the isotope mutated the DNA inside their sperm cells, so when the flies later hatched they came out firing blanks.

After the sterile insects successfully cleared an island test site of screwworms, irradiated pupae were deployed over mainland America. The first entomological air raid was in 1958 and after that, fresh pupae were dropped regularly. When the project was in full swing, the rearing facility in Sebring, Florida churned out more than 50 million sterile screwworms a week. The plan worked like a dream. By early 1959, the screwworm had vanished from the entire south-east United States. After that, the government started dropping the flies across Texas and the south-west, and by 1966 the whole of the US was screwworm free. It was a remarkable achievement; but more was to come.

Concerned that screwworm flies from Mexico and Central America could re-infest the US, in the following decades authorities air-dropped sterile flies into these regions too and little by little, the screwworm succumbed. By 1997, the screwworm had been wiped out all the way from Texas down to Panama.

If bombing fields with irradiated insect pupae sounds like science fiction, you may be surprised to learn that the process is still ongoing. There are still plenty of screwworms in South America and people are worried they could work their way northwards and re-infest the United States via Central America and Mexico. So a programme maintained by the United States and Panamanian governments still regularly deploys the insects over eastern Panama and parts of Columbia. A specialised production facility in Panama employs around 400 people to mass-produce the insects and so far the programme is working well. The sterile flies create a barrier that prevents re-infestation in the US.

'The whole programme has been extraordinarily successful,' says Luke Alphey from the UK's Pirbright Institute, who works in pest control. Unlike pesticides, which often kill the pest and the surrounding wildlife, the sterile insect technique kills only those insects it is designed to kill. Male screwworm flies only mate with female screwworm flies, so there is no damage to other species. And unlike a pesticide, which stays where it is put, the sterile insects disperse and actively look for their quarry. 'It not only eradicated the pest on a continental scale, it has also managed to maintain that eradication,' says Luke. The United Nations called it one of the 'greatest achievements in animal health' in the twentieth century.

Parasitic Planet

We influence evolution when we domesticate, selectively breed, genetically modify and clone organisms, but we also shape the broader evolutionary picture when we kill organisms en masse. There are millions of parasitic species. Most groups evolved before the Mesozoic era, more than 250 million years ago, then diversified and specialised to feed on the bountiful life forms that continued to emerge. When our ancestors emerged in Africa, they were plagued by biting and stinging creatures, but it's only in our recent history that humans have begun to quash these pests on any meaningful sort of scale.* Until recently, chemical pesticides have been the

* Insecticides kill just insects, while pesticides kill many different sorts of pests. The first known pesticide was elemental sulphur. The ancient Sumerians used it to protect their crops around 4,500 years ago. The first insecticide was probably nicotine sulphate. It was extracted from tobacco leaves and used in the seventeenth century.

front line, but now scientists are developing more sophisticated methods that harness the power of genetics. Our new tools give us the power not just to destroy local populations of pests; they potentially give us the power to wipe entire species from the face of the Earth. This is evolutionary warfare. Now, as our global dominance reaches new levels, society needs to decide how, when and if these tools should be deployed.

The US Department of Agriculture's Screwworm Program is a key player in this story. Forget altering the genomes of individual animals, this project altered the genomes of thousands of animals all at once. In the southern US and Central America, it has successfully managed to crush a parasite that has been plaguing livestock for thousands of years. The technique has since been rolled out to other parts of the world to deal with different parasitic infections. The sterile insect method has helped rid Zanzibar of the tsetse flies that cause nagana, a chronic disease of livestock, and the southwestern United States of the pink bollworm that feasts on cotton plants.

Although the insects may not have had their DNA precisely edited by CRISPR or some other modern molecular method, they have still had their genomes altered. Radiation introduces random changes to the DNA of living things, so the approach is scattergun but effective. The hard part is getting the right dose. Too little and the radiation has no discernible effect, but too much and the insects become incapacitated on multiple levels. As well as becoming sterile, they might become worse at flying or feeding or having sex. In addition, the dose that works for one species doesn't necessarily work for another. 'It made me realise we must be able to do better,' says Luke Alphey.

So 25 years ago, Luke came up with a new spin on the old method. Instead of using radiation to sterilise insects, he decided to use genetics, and instead of focusing on screwworms, he turned his attention to mosquitoes.

Mosquitoes are the most dangerous animal in the world. The blood-sucking parasites harbour disease-causing viruses and other parasites, which they pass on to humans and other animals when the females bite.* Mosquitoes are responsible for 17 per cent of the global burden of infectious disease, including every single case of malaria, Zika virus, yellow fever, West Nile virus, dengue fever and Chikungunya. Every year, they kill more than 700,000 people, which is equivalent to wiping out the entire population of Leeds or Denver.

Luke imagined adding in a gene that would cause the mosquitoes' offspring to die young, before they could reproduce. To generate multiple generations and the large numbers needed for pest control, the insects would need to be reared in some sort of production facility, meaning that sometimes the gene would need to be turned off so that they could reproduce successfully. So Luke envisaged adding in a chemical switch. When the insects were reared in the lab, a chemical added to their diet would turn the toxic gene off, but when the insects were released into the wild, where the chemical was absent, the gene would be unleashed to do its work.

After proving that his system worked in fruit flies, he next added it into *Aedes aegypti*, a particularly unpleasant

---

* Male mosquitoes do not bite.

mosquito species that transmits multiple diseases, including dengue, yellow fever and Zika. The technology was licensed to Oxitec, a British biotech firm co-founded by Luke. They subsequently developed the 'Friendly™ Mosquito,' a GM-engineered sterile insect that can be released into the wild.

When regulators were deliberating over the future of the GM AquAdvantage Salmon (see Chapter 3), they worried that if the fish escaped and bred with its wild relatives, its eclectic genes would end up tainting the wild gene pool. This was a reasonable concern for the salmon, but it's a different story for the Oxitec mosquitoes. They are designed to breed with the wild population but because the insects are sterile, they cannot spread their genes beyond a single generation. The sterile mosquitoes that are released die after a few days and their offspring die prematurely before they can pass their genes on. It's classic GM technology, but it's also deliberately short-lived. To make it safer still, the process has been tweaked to include a fluorescent marker gene so the insects can be tracked, *and* so that only sterile males are produced. Females, remember, are the ones that bite, so no one wants extra females – genetically modified or otherwise – to be set free.

Working with local governments, Oxitec has run various field trials in the Cayman Islands, Panama, Malaysia and Brazil, and the results are promising. After the insects were released into a suburb of Piracicaba in Brazil in 2016, wild populations of *Aedes aegypti* fell by more than 80 per cent. This is a big improvement on more traditional anti-mozzie methods like insecticide sprays, and one year later, the population remained similarly reduced. Now Oxitec finds itself somewhere in the fuzzy hinterland between promising field trials and commercial acceptability. In Brazil, for example, the country's national biosafety body, CTNBio, declared the mosquitoes safe for commercial release, but Anvisa, the regulatory body overseeing the trials, said the company needs to prove that its technology does more than eliminate the local mosquitoes. It's all very well showing that they 'drop like flies' but Oxitec now needs to prove that disease rates are declining as a consequence.

The Oxitec mosquitoes are a modern genetic twist on a tried and tested idea: Knipling's sterile insect technique. Instead of being chucked from a plane, the 'friendly' mosquitoes are often liberated from containers on the side of the road. Instead of being sterilised with radiation, they are sterilised with genetics, but the end product is the same: large numbers of sterile insects that infiltrate the wild population, causing it to crash, and then die out themselves. The only problem is that it's virtually impossible for the sterile insects to seek out and mate with every single last member of the wild population in one release, so the sterile insects need to be mass-produced and released on a regular basis. This is expensive. It costs the US and Panamanian governments US$15 million (£12 million) a year to produce and deploy the sterile screwworms needed to prevent their South American relatives from re-infesting America. Nor does the approach work for every disease. Dengue is predominantly a disease of urban spaces, so it's possible to imagine inundating whole cities or parts of cities with sterile mosquitoes concentrated in a relatively small area. Malaria, on the other hand, is more of a rural disease.

It is spread by a different type of mosquito, *Anopheles*, that is widespread across much of sub-Saharan Africa, making the sterile-insect approach unfeasible, so scientists have been developing alternative methods to try to control the world's *Anopheles* mosquitoes.

Serial Killer

Malaria is a serial killer. Every year, hundreds of thousands of people die after becoming infected with the *Plasmodium* parasite that the *Anopheles* mosquitoes carry. Most of these are children under five, and most live in Africa. Many millions more survive infection but suffer a painful and debilitating illness that can leave them weakened and more susceptible to other diseases. Nearly half of the world's population lives in malaria-prone regions and currently, at any one time, around 3 per cent of people on Earth are infected. Despite global initiatives to curb the disease using insecticides, repellents, bed nets and drugs, malaria remains one of the most severe public health threats in the world. In Africa, a child dies from malaria every minute.

Around 50 years ago, researchers began to wonder if they could eliminate malaria and other mosquito-borne diseases by repurposing a naturally occurring genetic phenomenon. In sexually reproducing species, most genes have a 50 per cent chance of being inherited by each of the offspring: these genes can either come from Mum or they can come from Dad. This is classic Mendelian inheritance, as laid out by the pea-growing monk, Gregor Mendel, in the nineteenth century. But some genes, known as selfish genes, bias inheritance so they are inherited more often – up to 100 per cent of the time. This means that selfish genes can spread through a population very quickly. Selfish genes, or remnants of them, have been found in many different sexually reproducing species, prompting scientists to wonder if they could be used to drive other, more useful changes through mosquitoes.

Suppose researchers were able to edit mosquitoes so that they could no longer be infected by the malaria-causing *Plasmodium* parasite, and then spread that change through the wild. Suppose the scientists could alter a gene so the flies were sterile, their eggs hatched out male, or the emergent flies had no wings. If the changes could be propagated through the wild population, then the mosquitoes would eventually die out.

It's a concept that has come to be known as 'gene drive'. Gene drives are selfish DNA sequences that can be used to spread desirable characteristics through populations of animals and plants. In the 1960s, researchers recognised their potential but didn't know how to make them. 'There were ideas in the air,' says evolutionary geneticist Austin Burt from London's Imperial College, but the molecular techniques needed to make them happen had yet to emerge. Then, in 2003, Austin suggested that homing endonucleases, a naturally occurring type of selfish gene, could be used to help build artificial gene drives. Homing endonucleases are enzymes that cut DNA and in theory can be used to help introduce the precise genetic changes that are needed to make a gene drive work, but they were finicky and difficult to work with. As new classes of DNA-cutting enzymes began to emerge the technology started to progress, but then CRISPR came along and everything changed.

When paired with a guide molecule, such as the enzyme Cas9, CRISPR can be used to make precise changes to the DNA of living things. It's versatile, easy to use and has been proven to work in just about every species that researchers have looked at. In 2014, two years after CRISPR first emerged, evolutionary biologist Kevin Esvelt – now at the MIT Media Lab – and colleagues wrote a paper describing how it could be used to make the sort of gene drive that scientists had been dreaming of for decades. The system involves using CRISPR to edit the gene of choice *and* paste in the instructions for making more edits, so when the DNA is inherited the process starts all over again.

At the same time, Kevin imagined the bigger picture of what could be achieved. Gene drive is a technology that takes genome editing out of the lab and into the wild. The idea is that when gene drive-edited organisms are set free, they will reproduce with their normal wild counterparts and pass a deliberate, engineered change to all of their offspring ... who will definitely pass it on to all of their offspring ... who will definitely pass it on to all of their offspring and so on. Applied to mosquitoes, it could be used to control or even eliminate diseases like malaria, dengue and yellow fever. Applied to other insects, it could be used to rid the world of Lyme disease, trypanosomiasis (African sleeping sickness) and others. In agriculture, it could be used to reverse rising levels of pesticide and herbicide resistance in insects and weeds, and in the wider world, it could be used to help control invasive species. Then a year later, Kevin and colleagues proved that the idea worked, when they built a synthetic CRISPR gene drive in the model organism, yeast.

Meanwhile, CRISPR's potential had not gone unnoticed by researchers working in the malaria field who were already starting to build gene drives in mosquitoes. Austin Burt and his colleague Andrea Crisanti work with Target Malaria, a not-for-profit research consortium trying to bring the disease under control. There are different sorts of gene drive, but Austin and Andrea are building suppression drives, so-called because their goal is to massively reduce or suppress the number of malaria-transmitting mosquitoes that exist. One of the drives – still in development – involves a gene that has been edited to bias the sex ratio, so the mosquitoes' offspring will all be born male. The other – well under way – targets a gene called *doublesex*, which determines whether an individual mosquito develops into a male or a female. The team altered the region of the *doublesex* gene that is responsible for female development, so when females inherit two copies, they are sterile. Males with the modified gene and females with one copy can reproduce as normal, providing a way to spread the infertility gene across generations. The result is that every generation, fewer fertile females are produced and fewer eggs are laid, so the population begins to dwindle. They tested their gene drive by releasing 150 edited mosquitoes into a cage containing 450 normal mosquitoes, and found that after just seven to eleven generations, the entire population crashed.

This is exciting. It's the first time ever that researchers have used a high-tech molecular method to completely block the reproductive capacity of a complex living organism. An important caveat: Austin and his colleagues are not trying to kill *all* mosquitoes. There are around

3,500 species of mosquito, of which only 40 related species carry malaria, all in the genus *Anopheles*. Target Malaria are setting their sights on three of these: *Anopheles gambiae, Anopheles coluzzii and Anopheles arabiensis*, because these closely related species are responsible for most of Africa's malaria transmission. Similarly, in the US, scientists are focusing on another related species, *Anopheles stephensi*, because it spreads malaria in India, the Middle East and South Asia.

There is a lot more basic research to be done, but the end game, says Austin, will be to release a few hundred mosquitoes into a few hundred villages in sub-Saharan Africa, and then leave them to get on with it … but there's a large ecological elephant in the room. It's already been shown that gene drives can be used to wipe out discrete populations of *Anopheles gambiae* in a lab setting. These gene drives are self-propagating. Released into the wild, there's a chance they could spread and wipe out the entire species.

Personally, I don't think the world would miss *Anopheles gambiae* or any of the other malaria-spreading mozzies. They have little ecological significance. They are not the sole food source of any known predator. The ecosystem would not collapse in their absence, but human lives would be saved.

I asked Austin if a world without malaria-transmitting mosquitoes would be such a bad thing. 'I guess I'm agnostic on that,' he replied. 'It's not our goal to eradicate them, but we would like to achieve a very substantial suppression. Ninety-five to ninety-eight per cent would be good.' This, he feels, is a reasonable goal. Africa is a very large place, and there are lots of mosquitoes in it.

'I think it's unlikely we would completely eliminate *Anopheles gambiae*,' he added, 'but I'm not sure I could guarantee, at this point, that it won't happen.'

The stakes are high. Humans have been exterminating species for millennia but this has largely been by accident, via over-exploitation or environmental change. Aside from a couple of notable exceptions,* it's actually very rare for humans to go out and deliberately kill every single last member of a living species. We're unintentional exterminators, not deliberate mass murderers, and yet, here is an insect that kills more people than any other animal. Surely gene drives are worth considering.

Suicide Possums

The Australian brushtail possum is a striking beast. With its pointy ears, protruding eyes and thick bushy tail, it bears more than a passing resemblance to the Pokémon character, Pikachu. It was first introduced to New Zealand in 1837 by settlers who were trying to establish a fur industry, but the unfortunate animals did not survive. If only they had stopped there. The traders kept on trying and in 1858 they managed to establish New Zealand's first wild possum population, in the furthermost tip of the country's South Island. After that, more animals were introduced and by 1930, they had been liberated in 450 different locations across New Zealand.

The possums settled right in. New Zealand has a rich and varied landscape. Its flora and fauna have evolved in

* Thanks to our actions, the world is now free of Smallpox virus and Rinderpest virus. Attempts to exterminate the Guinea Worm are currently under way.

a world bereft of land-living mammals or marsupials, and so in time, birds came to predominate. These, however, didn't stand a chance against the possums. The prodigious scavengers are not fussy eaters. They ate the eggs and chicks of native birds like kōkako and kiwi. By eating fruits and flowers, they made life difficult for other bird species like tūī and kākā, which depend on the high-energy food during their breeding season. They attacked the country's native bats and feasted on the islands' snails and invertebrates. A single possum can eat 60 *Powelliphanta* snails in a single night. They stripped trees and caused forests to die back, converting ancient closed woodland into open shrubland.

Now there are around 30 million possums living off the fat of New Zealand's land, where they outnumber the human residents by six to one. As well as their continued disregard for wildlife, they also lay into orchards, plantations and conservation plantings, and threaten the country's livestock industry by acting as a reservoir for bovine tuberculosis. In short, possums have become a total nightmare.

Invasive species, like the possum, have become a signature of humanity's global domination. Wherever we have gone, other species have followed in our wake. Possums were introduced to New Zealand intentionally, as were the pigs, deer, goats, sheep, rabbits and stoats that now thrive there. Early settlers accidentally introduced rats, and now species hitch rides on the soles of our shoes and the underbellies of our vehicles. Animals and plants transferred on aeroplanes are being transported at hundreds of miles an hour. We have vastly accelerated the rate at which species are travelling between continents, and as a consequence clashes between native and non-native species are increasing. 'We can be fairly confident that there has never been a time when so many species from geographically separate parts of the world have become mixed up so quickly,' says ecologist Chris Thomas from the University of York.

Although many invasive species present little threat to their new environment (see Chapter 7), others cause chaos. Invasive species have caused around 60 per cent of recent bird, mammal and reptile extinctions, and they continue to put many hundreds of species at risk. 'Invasive species are probably the largest single cause of vertebrate extinction since humans spilled out of Africa,' Chris says. Island-dwelling species are particularly vulnerable because they are hemmed in by water and – like the birds of New Zealand – lack natural defences against the invaders. Accordingly, island-dwellers account for 81 per cent of the species at risk from invasive predators, and non-native species are widely recognised as a critical threat to island biodiversity.

New Zealand has lost more than 50 species of endemic birds to invasive species, and today these non-native animals continue to kill around 25 million birds a year. Desperate times call for desperate measures. In 2016, the then Prime Minister John Key announced an audacious plan to eliminate every single last one of the country's invasive vertebrate predators – including brushtail possums, rats and stoats – over the following 34 years. He called it Predator Free 2050.

If it sounds like a tall order – which it probably is – then you have to remember that around the world, more than 1,000 islands have already been cleared of invasive species in so-called 'mega-eradication' programmes. New Zealand, which boasts some of the leading experts in the field, has carried out more than 200 of these. New Zealanders are world leaders in killing rats and stoats. Every year, New Zealand spends more than NZ$70 million (£37 million) on pest control. The exterminators have an arsenal of weapons that includes sophisticated trapping techniques, guns, and poisons like the controversial 1080, which has been used to help control New Zealand's pests for over 60 years. Deployed from helicopters, it is the cheapest and most effective way of dealing with the problem, but environmental groups don't like it. While the supposedly 'mammalian' poison has successfully wiped out vast numbers of rabbits, stoats and possums, it also kills birds like the endangered kea and game animals like pigs and deer. The sheer scale of the project also creates a problem. The largest island cleared to date, Macquarie Island in the south-west Pacific, measures 128 square kilometres (31,600 acres). New Zealand is 2,000 times bigger, and full of forests and cities and people's gardens. There are plenty of places for predators to hide away, so while it's currently possible to suppress New Zealand's pest population, it's not currently possible to exterminate them all.

New methods are needed for Predator Free 2050 to reach its goal, so now scientists are talking about 'next generation' pest control. Gene drive is one of those options. Imagine using CRISPR to alter a gene that is needed for reproduction – just as Target Malaria are doing for *Anopheles* – and then forcing that gene through all of New Zealand's possums using a gene drive. Small groups of 'suicide possums'* released in strategic locations would then disperse and breed with their wild roguish counterparts. Pretty soon the possum population would be no more.

I'm intrigued by this proposal for many reasons. New Zealand is riddled with pests. Suicide possums are just one suggestion, but one could equally imagine suicide rats, mice, stoats and invertebrates. Compared with poisons and traps, this is humane. The GM animals would live normal lives but their reproduction would be compromised. The approach doesn't involve toxic chemicals and appears to be species specific. Unlike using pesticides such as 1080, there should be no collateral damage to non-targeted species. It's cost effective because, for example, a single release of suicide possums should be enough to do the job, *and* it's permanent. The gene drive would, in theory, keep on going until the last possum standing finally bites the dust.

Gene drives have the potential to spread a particular genetic mutation through an entire population or species. This makes them both enticing and terrifying in equal measure. It's questionable whether anyone would mourn the passing of *Anopheles gambiae* in its entirety, but suppose the suicide possums spread. Brushtail possums are vilified in New Zealand, but adored in their native Australia, where they are protected. If the suicide possums spread to Australia, they *could* end up exterminating all of the native population.

It is early days. Gene drives are still very much in their infancy. They may well have been demonstrated in

---

* Suicide Possums sounds like the name of an intense Indie rock band.

mosquitoes, fruit flies and yeast, but their transition to the vertebrate world is likely to prove challenging. In 2019, Kimberly Cooper from the University of California in San Diego managed to make gene-drive mice – not for eradicating invasive species, but in order to create better lab models of disease – but found, inexplicably, that the procedure only worked in female animals. A gene drive for possums could prove even more problematic. Being a marsupial, their reproductive biology is vastly different from that of mice. Their genomics is poorly understood and to date, no one has managed to create a genetically modified possum of any description. So we're not about to see a suicide possum any time soon.

Dangerous Driving?

When Kevin Esvelt first suggested building CRISPR gene drives back in 2014, he proposed their potential use to help curb invasive species. Then three years later, he did a massive U-turn. In a paper written for *PLOS Biology*, Kevin and his colleague Neil Gemmell outlined all the reasons why this could be a very bad idea. As the gene-drive animals will be around in the wider environment for a while before they die out, it provides them with a window of opportunity for escape and proliferation. 'Any extended residence time provides an opportunity for them to hitch a ride to other islands and continents,' he says. Rats and mice, for example, are sturdy ocean wayfarers well known for their ability to stow away on boats or drift on chunks of floating debris. They have already spread around the world once. There's every reason to expect they could do it again, but if they carried a gene drive designed to kill their kind, it could spell the end of every last rat on Earth.

Even if the rats did not stow away of their own accord, there's always the chance they could be deliberately relocated. There's certainly a precedent for this. In 1997, a clandestine group of farmers illegally smuggled the calicivirus responsible for Rabbit Haemorrhagic Disease into New Zealand. Rabbits – a non-native species – had reached pest proportions and the farmers were disgruntled with the perceived lack of governmental activity. So they took matters into their own hands.

In the United States alone, rats cause economic damage estimated at US$19 billion (£15 billion) a year. 'If I were a poultry breeder, I'd be stupid not to hire a mercenary to go out and bring the gene-drive rats back to my farm,' says Kevin. 'The notion that people would not mess with your trial and move your gene-drive rats is zero.' The bottom line, he says, is that creating a self-propagating CRISPR gene drive is little different to creating a new, highly invasive species. Both are likely to spread and both have the potential to cause serious ecological damage. Once the gene-y is out of the bottle, it could be very hard to get it back in again.

Even making the organisms could be risky. Fruit flies, for example, are a commonly studied model organism, and it's not uncommon for the odd one to escape from the lab. The first gene-drive fruit flies were made back in 2015 in a proof of principle experiment that saw an edited gene for yellow pigmentation spread rapidly through the population. None of the flies escaped from the lab, but it's been estimated that if one had, by now up to half of all the world's fruit flies would be honey-coloured. In time there wouldn't be a single brown one left.

It's for all these reasons that Kevin thinks it would be unwise to release a self-propagating gene drive that can spread indefinitely *unless* the explicit goal is to target an entire species. 'I have nightmares about yellow fruit flies,' he tells me. 'If I develop a medicine and your doctor recommends that you take it, you can say no. But if I design a gene drive designed to alter the shared environment where you live, and you vote against it and you're outvoted, then you're going to be affected, no matter what. This kind of technology is ethically very different. If you are going to release something that will spread on its own, you need to make sure you have the permission of every country that harbours that species.' Just imagine how easy that would be.

Like many other people, the residents of New Zealand have mixed feelings about GM. While some embrace the prospect of possible technological fixes, others feel a deep-seated sense of unease at the prospect of 'meddling' with nature. New Zealand farmers don't grow GM crops, and some regional and city councils are busy implementing anti-GM legislation. This will make it difficult if the country ever seeks to adopt a nationwide policy. Further afield, environmental groups, like Friends of the Earth and the ETC Group, have repeatedly lobbied for a temporary ban on the release of gene-drive organisms while their merits are debated, but so far their proposals have been rejected. As it stands, the UN Convention on Biodiversity – an international treaty that seeks to conserve biodiversity – has ruled that scientists should assess the risks of gene-drive releases on a case-by-case basis, and consult with any local communities and indigenous groups that could potentially be affected.

Caution, communication and transparency are vital. More than any other technology, gene drive has the power to rapidly skew evolution and send entire species careering along entirely novel courses. Natural selection famously favours the survival of the fittest. The genes that hang around are the ones that help organisms to survive and reproduce, but gene drives don't play by the rules. Synthetic gene drives are unique because they can be used to spread both beneficial and harmful genes through a population. This is not a technology to be unleashed lightly.

Kevin believes that it calls for a shift away from the classic scientific paradigm where scientists receive regulatory approval, perform their experiments and *then* tell the world what they have found. Because gene drives have the potential to affect shared environments, Kevin thinks that scientists should be talking to people early on, when any experiment or intervention is at its conceptual stage. 'It's the decisions that we make early on in the development of a technology that have the most influence on what the eventual application is,' he says. If people are dead against a particular technology, then scientists must pay heed and stand down. If they are broadly in agreement but have specific concerns, then the issues need addressing. In this paradigm, the public would help to design the sorts of experiments that people like Kevin do. 'I view it as unethical to develop these kinds of technologies without telling the world what we plan to do and inviting their concerns, criticisms and suggestions for how we can do it better,' he says.

It's an honourable goal, but there's just one problem. People need information. They need to know how the release of a gene drive might affect them and the things they care about. What are the benefits? What are the risks? Computer modelling can only go so far, so at some stage experiments must be done in order to generate the information needed to make evidence-based decisions.

Peter Dearden is the Director of Genomics Aotearoa, a New Zealand-based initiative that seeks to harness the power of genomics to improve human health, primary production and the environment. In some shape or form, genomics will be part of Predator Free 2050. Studying the genomes of possums and stoats, for example, could lead to the development of species-specific poisons. Gene drives *could* prove useful. 'What worries me,' says Peter, 'is that if the general feeling is that the public doesn't want to go for GM or gene drives in the absence of data, that that will have a chilling effect on the science. Maybe it never goes ahead and we never get to the point of being able to present the risks and the benefits.' But how can scientists ever run contained field trials of a technology that is designed to keep on spreading? It's a classic Catch-22.

There are several answers, but the first is to start small. For pest control, Peter imagines starting with contained, controllable experiments. In the early stages, these would be in a laboratory, then if the public agree, they could be moved to isolated external test sites. New Zealand is home to more than 600 islands, so perhaps one could be set aside as a living laboratory. The gene drive could be tested in relative safety with limited opportunity for its accidental spread.

One day Target Malaria would like to test its gene-drive mosquitoes in the wild. The scientists behind it have shown that their project works on insects in a caged laboratory setting, so wild release is the next obvious step. There is no exact precedent for this, but remember that Oxitec have already released regular GM mosquitoes that are sterile into test sites in a handful of different countries. Target Malaria is planning to do something similar. In 2018, the government of Burkina Faso granted permission for them to release up to 10,000 male GM sterile mosquitoes. They will be genetically modified but they won't contain a gene drive, and will quickly die out after mating with wild females. By Target Malaria's own admission, it's unlikely to make much of a dint in the continent's mosquito population. Instead it's about testing the societal waters and acquiring incremental knowledge. Local scientists will gain first-hand experience of raising and releasing GM mosquitoes, while local communities will hopefully learn that the insects are not to be feared. Target Malaria is also working with partners in Mali and Uganda, where similar trials could be envisaged. It's a step-by-step approach that involves working with local communities and regulators to ensure acceptance and approval at each stage.

If all goes well, at some point, the next step will be to release a small cadre of gene-drive mosquitoes. 'We can expect them to be crossing borders,' says Target Malaria's Austin Burt. A gene drive released in Burkina Faso, for example, could spread through the whole of the continent, so Target Malaria will only proceed if they have full backing from all of Africa's regulatory bodies. 'It's not going to happen tomorrow,' says Austin, 'but we can imagine putting in a dossier asking for permission to release the insects in one or two villages as a test in 2024.'

Scientists should start small, tread lightly, act transparently and listen to people. Sensitive to concerns of their untrammelled spread, researchers are also devising methods to make gene drives more manageable. If a gene drive were to spread too far, for example, scientists could choose to release organisms containing a second drive that overwrites the instructions of the first. It would be like an antidote to a toxin. Another option is to design a drive that naturally runs out of 'oomph'. Kevin Esvelt is developing 'daisy drives' – so-called because they depend on separate parts like a daisy chain – which are designed to work for a while, then stop. Alternatively, they could use 'threshold drives', which only spread if their frequency exceeds a particular threshold in the population. If such a drive ever spread into a nearby population, it would be outnumbered by the wild organisms it encountered and then naturally grind to a halt.

In the end, scientists may go to all this trouble and then find that the gene drives they make don't even work that well. Charles Darwin once described natural selection as 'a power incessantly ready for action', as 'immeasurably superior to man's feeble efforts, as the works of Nature are to those of Art'.* When you consider the long game, natural selection has a pretty good track record of booting out mutations that don't do life any favours, and in lab experiments, researchers have noted the rise of resistance. Mosquitoes can become resistant to

* Clearly Darwin was no Art buff!

gene drives just as insects can become resistant to pesticides, although the extent of this phenomenon is as yet unknown. Natural genetic variation may also pose a problem. CRISPR gene drives work by recognising short DNA sequences, but if individuals differ at these sites then the gene drives may not work. A recent study of *Anopheles* mosquitoes taken from across Africa highlighted extreme genetic diversity, suggesting that the number of potential gene-drive targets may be limited.

If they do work, it's likely they will work alongside other approaches. Target Malaria is unlikely to do away with chemical sprays and bed nets, any more than the scientists involved in Predator Free 2050 are likely to stop using traps or pesticides. We may find it's actually quite difficult to drive these pest species to extinction. Oh, the irony! Humans spend thousands of years accidentally driving species to extinction, only to find that when they decide to do it deliberately, it's much harder than they thought.

When I pressed Kevin on his reservations surrounding their release, he told me that the ecological repercussions don't actually concern him. 'It's not a physical or ecological threat,' he says. If scientists are smart enough to design gene drives, they are almost certainly smart enough to design the technology needed to keep them in check. Kevin thinks it will be possible to put the gene-y back in the bottle. Instead, he is concerned that the premature release of a gene drive – say in New Zealand – could spark an international incident that erodes public confidence in an already tainted and politically charged arena. The field of gene therapy is still rebounding from the tragic death of Jesse Gelsinger, an

18-year-old who died in 1999 after a gene-therapy trial he took part in went wrong. The unauthorised release of a gene drive into the wild could quite likely set the field back by a decade or more.

Sixty years ago, we reined in populations of unwelcome pests by airdropping millions of irradiated pupae onto the soils of southern America. Now, if gene drives are proven to be safe, acceptable and effective, we may choose to control pests by releasing just a few well-chosen genetically altered individuals. This could bring improvements for human health, ecological health and agricultural health, but it's currently something of an unknown quantity. We won't know the long-term consequences of releasing a gene drive into the wild until we release a gene drive into the wild. I'm in favour of developing the technology at least to the point where we can make a genuine assessment of the related risks and benefits, and then decide what to do. The regulators won't move quickly on this. They never do. So on the one hand we have time, but on the other hand invasive species continue to wreak havoc, and every day malaria kills between 1,200 and 2,000 people. Sometimes the cost of activity can be high, but sometimes the cost of inactivity can be even higher.

CHAPTER SIX

The Age of the Chicken

Over the last 250 years, scientists have divvied up the Earth's epic 4.5 billion-year-long history into smaller, more manageable chunks. Eons, eras, periods, epochs and ages are all discrete slices of geological time, nestled inside each other like Russian dolls. Eons – the biggest timespans – contain eras which contain periods which sometimes contain epochs and within them, ages.

Around 20 years ago, atmospheric chemist Paul Crutzen was sitting in a scientific conference in Cuernavaca, Mexico, where researchers were discussing the events that had shaped the most recent epoch, known as the Holocene. The Holocene began around 11,700 years ago when giant ice sheets retreated and the Earth finally shrugged off the last Ice Age. The world warmed and tundra gave way to forest. Life became less of a struggle as people began to domesticate farm animals and plants. Agriculture blossomed, cities sprang up and the human population began to grow. All of our written history comes from this time, but sitting in the meeting, Crutzen found himself wondering if the Holocene had come to an end. It struck him that the changes currently being wrought on the planet are quite different from those that occurred during the main part of the Holocene epoch. Put simply, the world had changed too much for it still to be in the Holocene.

He blurted out, 'Let's stop it. We are no longer in the Holocene. We are in the Anthropocene …' The room fell silent. Scientific tumbleweed rolled across the floor. Then when the session came to an end and those assembled spilled out for coffee, the 'Anthropocene' was all anyone could talk about. What had seemed obvious to Crutzen now began to consume the geological community.

Crutzen's outburst was not premeditated or planned in any way. He later told a BBC journalist that he made up the word 'Anthropocene' on the spot. It was a good choice. 'Anthropos' is Greek for human, and 'cene' derives from the word for 'new'. The Anthropocene, Crutzen would later propose, stands for a new chapter in Earth's history: one where humans have become the major force driving planetary change.

For most of its history, our planet and its inhabitants have been shaped by non-human forces, such as the waxing and waning of ice sheets and global swings in temperature. Evolution was tempered by natural forces but now humans have become so numerous, so industrious and so blasé, that the repercussions of our collective actions are global. Now all life on Earth – I repeat – *all* life on Earth is influenced by the actions of a single living species: *Homo sapiens*, the so-called 'wise' people.* To this point, I've mainly discussed the

* If one were to be picky, it would be worth noting that humans are not, in fact, the first species to induce planetary-scale change. Over 2.4 billion years ago, during the Proterozoic Eon, single-celled organisms called cyanobacteria evolved and began using light to generate chemical energy. Oxygen was a by-product and in time it

evolutionary changes that humans have caused intentionally through processes like selective breeding and genetic modification, but in the next few chapters I'd like to consider some of the changes that humans make unintentionally.

When Crutzen was considering the Anthropocene, he was thinking about all the many ways that people have altered the planet. In recent times, we have razed forests and cleared vast swathes of natural land to make way for skyscrapers, shopping centres and industrial plants. Now roads, fences, canals, power lines and railways criss-cross the landscape. We have dammed and diverted rivers, altering the fate of the living things that they support. We have produced enough concrete to cover the surface of the Earth in a layer two millimetres thick. Underground, we have gouged minerals, metals and other natural resources from the bowels of our planet. Every year, human activities move more soil, rock and sediment than is transported by all other natural processes combined. In South Africa's Kalahari Desert, tree roots reach over 60 metres (200ft) down, yet gold miners in the same country have tunnelled 5,000 metres (16,400ft) underground. Meanwhile, in north-western Russia, the world's deepest borehole penetrates more than 12,000 metres (40,000ft) under the Earth's surface.

---

started to mount up in the atmosphere. Before this so-called Great Oxidation Event, life was dominated by anaerobic micro-organisms but as the atmosphere changed, the stage became set for the evolution of aerobic life forms, including eventually you and me. Cyanobacteria may be little, but they are mighty. They changed the future and face of life on Earth.

We fell deeply, madly, obsessively in love with plastic. Now our oceans are stuffed full of it. Some researchers estimate that in the next 30 years, there will be more plastic in our oceans than fish. Large pieces entangle marine animals. Smaller pieces are eaten by them. Ninety per cent of the world's seabirds now have plastic in their stomachs, and the substance has become so ubiquitous that tiny plastic particles can now be found in our drinking water and in some of the animals that we consume.

Our actions have disrupted the global cycling of elements that is needed to support life. We douse our crops in pesticides and nitrogen-rich fertilisers, which run off into the rivers and pollute the oceans. As if plastic wasn't enough to contend with, there are places where life-suffocating dead zones are strangling marine life. There are now more than 400 dead zones in the waters surrounding our coasts. The largest, in the Gulf of Oman, is almost the size of Florida and still growing. Life doesn't stand much of a chance in these oxygen-depleted waters.

Things are just as gloomy in our atmosphere. Since the Industrial Revolution, we have added 2.2 trillion metric tonnes of carbon dioxide into the atmosphere, boosting levels by more than a third. This has been caused by the burning of fossil fuels and by the mass destruction of carbon-absorbing forests. Now the atmospheric concentration of carbon dioxide is higher than at any time in the last 800,000 years.

This is making our oceans more acidic and raising the Earth's temperature. In a sense, climate change is nothing new. The Earth has gone through multiple periods of warming and cooling before. What's worrying is that this current period of warming is happening more quickly than many past events. Scientists are concerned that this rapid human-induced warming is outpacing the natural variation that occurred previously, with serious repercussions for the stability of our planet's climate.

Temperature records dating back to the late nineteenth century reveal that the Earth's surface has warmed by about 0.8°C in the last 100 years. This might not sound like much, but it's more change than occurred during the entire Holocene epoch, and it's having a big effect on the world around us. The weather is becoming wilder and less predictable. Glaciers are melting, sea levels are rising and homes and habitats are being destroyed by flooding. And don't even get me started on the repercussions for animal and plant life. We'll come to that shortly.

In October 2018, the Intergovernmental Panel on Climate Change (IPCC) released a report warning that, unless we radically rethink the way we live, the world is on track for around 3°C of warming by the end of the century. To put this in context, it's thought than an increase of 1.5°C relative to pre-industrial levels will already interfere with the planet's ability to support life. There will be more extreme weather, including severe heatwaves and intense storms. Things get worse as the figure increases. Two degrees of warming will increase the risk of drought, floods, extreme heat and poverty for hundreds of millions of people. Coral reefs could be entirely destroyed (see Chapter 9), as could 13 per cent of the world's land-based ecosystems. Beyond that, three degrees of warming is, frankly, not somewhere we want to go. When the report came out, Kaisa Kosonen, Senior Policy Advisor at Greenpeace Nordic, told the BBC, 'Scientists might want to write in capital letters, "ACT NOW, IDIOTS".'

If time travel were possible and a person could be transported from the start of the Holocene, 11,700 years ago, to this moment in time, they would look around and be baffled. Instead of aurochs and tarpan (the Eurasian wild horse) there would be modern cattle and Shetland ponies. Instead of open fires and stone tools, there would be waffle makers, electric ovens and fast-food outlets. The world that you and I take for granted would be almost entirely alien to them.

Crutzen was right. In a relatively small slice of geological time – the last 12,000 years or so – we have changed the Earth profoundly. Now it is smothered in the fingerprints of human activity. As we change our climate and warm our world, there is no corner of the air, land or sea that is left untouched. Humans have caused change on a planetary scale, and now scientists are debating whether the Anthropocene should be formally recognised.

While many believe that it should, scientific acceptance hinges on the discovery of a 'golden spike', an environmental marker chosen to epitomise the start of this proposed new epoch. It's the sort of thing that future geologists will look back to in millions of years' time and say 'There! That is it. That is the moment when the Anthropocene began.' When the trappings of today have become fossilised in stone, the Anthropocene's golden spike will be found in a single, specific place, in a particular layer of rock.

If, for example, you want to see the place that typifies the dinosaur-laden Cretaceous Period transitioning to the subsequent Paleogene Period, then take a trip to El Kef in Tunisia. Grab a rental car and head towards the Roman bath house at Hammam Mellegue then veer off-road near a large electricity pylon. The golden spike lies in a dusty hillside at 36.1537°N, 8.6486°E. If dinosaurs aren't your bag and you prefer to visit the place that best represents the blooming of complex life at the start of the Phanerozoic Eon, head to the Burin Peninsula in Newfoundland, Canada. The golden spike for this boundary can be found halfway down the Fortune Head rock outcrop at 47.0762°N, 55.8310°W.

Theoretically, if similar features are found in a similar sequence of rocks somewhere else in the world, then those rocks will be the same age. The locations are called golden spikes because they act as gold standards for rocks of a particular age and because early geologists used to literally mark them with metal spikes.

Sometimes it's the chemical composition of rocks that set them apart from their older and younger neighbours; other times it's the fossilised life forms they contain. Sometimes, it's both. The rust-coloured rock layer that threads its way through the hillsides of Tunisia, for example, is laced with unusually high levels of the element iridium, which found its way there when an iridium-rich asteroid crashed into the Earth and annihilated the dinosaurs. Meanwhile in Canada, fossilised worm burrows mark the moment when life exploded at the start of the Phanerozoic. Casts of the U-shaped burrows made by relatives of today's marine priapulid worms[*] are commonplace in this particular rock layer and mark the start of the Eon that we still live in today. But what marker will come to denote the start of the Anthropocene? And where will that golden spike be found?

When future geologists look back to the rocks, ice cores, lake sediments, tree rings, fossils and shells from the Anthropocene, they will detect a list of anomalies as long as your arm. There will be odd isotopes. Some will be from the fallout of nuclear explosions like Hiroshima and Nagasaki; others the unmistakable signature of fossil-fuel combustion. The rocks will contain strange new materials. They will be laced with human-made inventions such as concrete, asphalt, steel, aluminium and plastic. There will be weird new minerals. It took planet Earth more than 2 billion years to produce the 5,200-plus natural minerals that exist, but in the last 250 years, we have added a further 208 to the list. Some, like the silicon chips used in semiconductors, were created deliberately, while others were made by accident. The mineral tinnunculite for example, is only formed in *very* specific circumstances, when droppings from the European kestrel meet and mix with the hot gas from coal fires,[†] while calclacite occurs when minerals kept in

---

[*] Priapulid worms, which look like penises, are named after Priapus, the Greek God of male genitalia (yes, there was such a thing). A popular figure in Roman erotic art, Priapus is frequently depicted as having a painfully permanent and ridiculously enormous erection.

[†] This is one of my favourite facts, not least because it begs so many questions, including, 'HOW did they find that out?', 'Does it have to be *kestrel* poo?' and 'Can I get a brooch made of that?'

oak drawers in museums react with chemicals in the wood.

All of these changes will be visible to future scientists, who will see it as evidence that the Earth went through a period of turbulent upheaval circa now. Any one of them *could* be the Anthropocene's golden spike, but the most obvious change won't be in the rocks' chemical makeup. Instead it will be in the fossilised life forms they contain. Different rock layers, laid down at different times, contain different fossils that typify life from that era. Because new layers of rock are laid down on top of old, abseiling down a cliff face is like travelling back in time. Future geologists will notice that as the world transitioned from the Holocene into the Anthropocene, the composition of life changed radically.

All Change

Before the Anthropocene, the world was full of wild and wonderful creatures. Most notably, the world's megafauna were still with us. Towards the end of the Pleistocene epoch (which came before the Holocene), the Earth was full of big animals. There used to be a giant, hornless rhino called *Paraceratherium*. It was roughly as tall and as long as a double decker bus. There was *Palaeoloxodon*, an enormous straight-tusked pachyderm that was twice the size of today's African elephants. There was a marsupial the size of a VW Beetle, giant armadillos with maces on their tails and a ground sloth so tall it could have peered into a modern-day first-floor bedroom window. Add to that more 'familiar' animals such as woolly mammoths, sabre-toothed cats and dire wolves, and the Pleistocene was packed full of larger-than-life characters. And then, one day, they started to vanish. Many potential reasons have been given. Some blame climate change, others disease, but increasingly the finger seems to be pointing in just one direction: at humans and their hominin ancestors.

Felisa A. Smith from the University of New Mexico has studied how the size of mammals has changed across time. In a 2018 study she concluded that, whenever humans are around, the mammals that go extinct tend to be 100 to 1,000 times bigger than those that survive. This in itself isn't new. It's well known that, as our ancestors migrated out of Africa and arrived in new places, the megafauna that lived there started to disappear. But the new study shows that this story has repeated itself on every continent except Antarctica over at least the last 125,000 years. To put it simply, when humans are around, big animals die.

Our ancestors killed the megafauna because they were a generous meal and a rich source of protein, while smaller, less filling species managed to escape. This systematic extinction began in the Pleistocene, extended into the Holocene and continues to this day. Now the world's megafauna are but a shadow of their former rich, diverse selves. In North America the average body mass of land-based mammals decreased from 98kg (216lb) in pre-human times to 7.6kg (16.7lb) in post-human times, and a similar trend can be seen across the globe. If all the species that are currently at risk were actually to go extinct, this would further decrease the body size of mammals. The average body mass of land-living mammals would reach its lowest value at any time in the past 45 million years. How would the fossil record look then? Future geologists would see that in the run-up to the Anthropocene, land mammals shrank spectacularly, as animals the size of reindeer and bears were supplanted by animals the size of otters and spider monkeys.

Today, Africa is the megafauna's final stronghold; although as strongholds go, it's weak. Most of our big animals now live in countries that are resource-poor and conflict-stricken. In the 1930s, the continent boasted around 4 million African elephants. Now, there are around 400,000. In the sixties, there used to be 2,500 northern white rhinos in central Africa but now there are just two. The global giraffe population has fallen by 40 per cent in the last 30 years. Now fewer than 100,000 exist and the once abundant species was recently added to the International Union for the Conservation of Nature (IUCN)'s Red List of Threatened Species. If these giants continue to be persecuted at the current rate, they could be gone from the wild by the time my future grandchildren grow up. This is not the world I wish for them.

Species are slipping through our fingers. Studies have shown that extinction rates are currently a thousand times higher than during pre-human times, and that it's not just the megafauna that are vanishing. In my insignificant lifetime, wildlife has taken a global beating. Now 25 per cent of mammals, 14 per cent of birds and 40 per cent of amphibians are in danger of extinction. We've lost 600 plant species from the wild in the last 250 years. In recent times, we have kissed a bittersweet goodbye to the Catarina pupfish (a tiny fish from Mexico), the Christmas Island pipistrelle (a bat that vanished from its island namesake) and the Bramble Cay melomys (an unassuming brown rat that lived on a tiny island in the Great Barrier Reef) – but these are just a few of the species that we know about. It's estimated that we lose between 30 and 150 species every day, but because most of these are invisible to us – hiding in faraway places, obscure, unmonitored or unknown to science – we remain in a state of torpid ignorance.

In 2017, scientists published a paper in the peer-reviewed journal *Proceedings of the National Academy of Sciences*. It was an overview that looked at how populations of 27,600 land-living vertebrate species – nearly half of all the ones that are known – had changed between 1900 and 2015. They found that a third of the species studied are in decline, and that even species not currently labelled as endangered are in trouble. When they looked at historical data for 177 mammalian species, they found additionally that every single species has lost at least 30 per cent of its range over the last century, and that half have lost more than 80 per cent. Ranges are shrinking and populations are dwindling. Lions, for example, used to roam all over Africa and could be found all the way from southern Europe to north-western India. Now, however, they exist only in scattered regions of sub-Saharan Africa and a single forest in India. In the last 20 years, their numbers have almost halved.

Let's be clear, the paper didn't focus on how many species were officially endangered or how many have gone extinct; rather it looked at a more subtle metric: the often invisible ebb of groups of animals over the course of time. Where extinctions mark the endpoint of loss, population declines describe the process of loss. Before a species vanishes completely, it first vanishes locally. This is important because each local loss is a stepping stone towards extinction.

In 2018, the World Wildlife Fund's Living Planet Report painted a similarly dismal picture when it reported that populations of mammals, birds, fish and reptiles have fallen by an average of 60 per cent since 1970. To put this in context, if there was a 60 per cent decline in the human population, it would be the same as emptying North America, South America, Africa, Europe, China and Oceania.

Now scientists are usually a pretty guarded bunch. They're not generally prone to alarmist statements; yet Gerardo Ceballos, Paul R. Ehrlich and Rodolfo Dirzo – the authors behind the 2017 paper – chose to describe the massive loss of wildlife that they observed as a 'biological annihilation'. Indeed, they felt the situation is now so drastic, it would be unethical *not* to use strong language.

Humans are responsible for the biggest dying the world has seen since the demise of the dinosaurs some 65 million years ago. Climate change, poaching, habitat loss, over-hunting, invasive species, disease and pollution all have their part to play. This is human-mediated evolution on a truly terrifying scale. The losses of today will become etched into the rocks of tomorrow, and future geologists may well wonder what the hell was going on. Vast amounts of biodiversity are being lost in geologically trivial amounts of time. Species are vanishing. Populations are waning. It's one thing when endangered species drop in number, but when abundant, seemingly bulletproof animals – like the barn swallow or the European hare or the grey mouse lemur – start to decline, it's quite another. It's not the start of a slippery slope. It's an indication that we're already halfway down the mountain. Birdsong is now conspicuously absent from my nearest town. The hum of crickets is gone from my local meadow. Life is diminuendo-ing from a glorious roar to a muted whimper.

Regardless of whether or not you consider yourself a nature lover, this is a global tragedy. It's no exaggeration to say that if things continue as they are this could be the unmaking of our world. When animals and plants disappear, so do the services they provide. All living things play a role in their resident ecosystem. They are pollinators, water purifiers, decomposers, seed dispersers, pest controllers and fertiliser-ers. They are dam builders and burrowers, carnivores and herbivores, predators and prey. We depend on the natural world for the air we breathe and every mouthful of food we eat. Wildlife isn't just a 'nice to have', it's essential. Human health, human homes, global food production and worldwide financial stability depend upon the natural world. If a price tag were needed to convince you, it's estimated that services relying on nature are worth around £100 trillion (US$125 trillion) globally. We must never forget that Earth's capacity to support life is shaped by life itself. Ceballos, Ehrlich and Dirzo called the situation 'a frightening assault on the foundations of human civilization' and cautioned that 'humanity will eventually pay a very high price for the decimation of the only assemblage of life that we know of in the universe'.

Future geologists will witness this whittling away of life in the rock layers that they study. The demise will become synonymous with the rise of the Anthropocene, so it could be used as a proxy for the epoch's arrival – the elusive golden spike. However, scientists are unlikely to characterise this new geological time period on the basis of lifeforms that went extinct. Instead, they're likely to be looking for a more overt and concrete measure. If life, rather than death, is to typify this new human-dominated epoch, then our future geologists will have to focus on one of the Anthropocene's success stories.

The Chicken of Tomorrow

As the world's megafauna were disappearing, another group of animals was rising to prominence. Ten thousand years ago, this group did not exist, but they are now so numerous that they make up 60 per cent of all the world's mammals and 70 per cent of all the world's birds. They are modern farm animals. There are currently 22 billion chickens on the planet. It's a staggering figure: enough for three per person. This makes chickens the most numerous bird in the world. They outstrip the most common wild bird species – the sparrow-sized red billed quelea of sub-Saharan Africa – by 14 to 1. After them, cattle are the world's most common large mammal, numbering around 1.4 billion, while there are around 1.2 billion sheep, 1 billion pigs and 1 billion goats. A recent study revealed that, while 96 per cent of all large, land-living mammals are either livestock or humans (60 per cent livestock, 36 per cent people), just 4 per cent are wild animals. The world is full of domesticated animals. Anthropocene rocks won't be full of fossilised wild species. They'll be full of fossilised chicken bones.

The chicken's story begins over 4,000 years ago in the Indus Valley of south Asia, where the birds were domesticated for the first time. The chicken's ancestor, the red jungle fowl, lived in the dense tropical forests of what is now modern-day Pakistan and India. It was, and still is, a shy and reclusive bird. Slender and athletic; males sport a livery of chestnut, marmalade and metallic green feathers. To the Bronze Age settlers who lived there, the birds must have seemed attractive. They didn't migrate as they were unable to fly long distances, so they were taken in to the settlers' villages and the pathway to domestication began.

Compared with other domestic stock of the day – cattle, goats, sheep and the like – chickens were easy to coop up and keep. They were readily portable, so when travellers set off in search of distant lands, they took their chickens with them. Phoenician traders introduced them to the Iberian Peninsula in the first century AD, and Spanish colonists introduced them to the New World in 1500. Yet as they spread around the globe, they changed comparatively little. The modern chicken is actually a very recent invention.

As recently as the end of the Second World War, chickens were still a scrawny affair. They were bred for egg production rather than for meat and when the birds *were* eaten it was either because they hatched out male, or because they were females that had stopped laying. Chicken meat was a luxury item, but as the population grew, breeders began to dream of an affordable broiler that grew quickly and could easily feed a family of four.

In 1945, the dream became the goal of a competition when the Atlantic and Pacific Tea Company challenged breeders to create the 'Chicken of Tomorrow'. The winners were announced at a fancy ceremony at the

University of Delaware's Agricultural Experiment Station, from a stage decorated thoughtfully … in chunks of frozen chicken meat. The triumphing birds were chosen because they grew rapidly to adulthood, and efficiently converted the food that they ate into meat on their bones. Henry Saglio, a teenager from Arbor Acres Farm, Connecticut, won best pure breed for a snow-coloured bird called the White Rock, while Californian brothers, Charles and Keith Vantress, won best hybrid for a cross between the New Hampshire and the Cornish Red. The Chickens of Tomorrow had arrived.

That evening, a parade passed through the streets of Georgetown, Delaware. There were balloons, flags and floats. Nancy McGee, a local girl, was given the unenviable title of 'Chicken of Tomorrow Queen' and she rode through crowded streets on a fur-covered chariot. The local chickens, I presume, ran for cover. Within a couple of years, the two winning strains were combined to produce a new crossbreed – the Arbor Acre – and in the decades that followed, selective breeding was used to evolve the bird into a giant.

Fast forward to today and our modern chickens weigh four to five times more than the broilers of 60 years ago. They put on weight around five times faster, and as a result are ready for slaughter in just five or six weeks: a 'saving' of around 10 weeks per bird. These are big, puffed-up birds with broad bodies and low-slung centres of gravity. Their rampant growth leaves its marks in their bones, which are less dense and often deformed. If they were left to grow old, they probably wouldn't survive. In one study, where the slaughter age was increased from five to nine weeks, there was a sevenfold increase in mortality rate. The rapid growth of leg and breast muscles that we have deliberately selected for leads to a relative decrease in the size of other organs like the heart and lungs, so the birds' physiology is compromised.

Today, more than 65 billion chickens are consumed every year and their bones are scattered across landfill sites and farms across the world. With numbers like this, there's a very good chance that modern chickens will be represented in the future fossil record, where their bones will stand out from the smaller, less deformed varieties that preceded them. Because the usually omnivorous birds are fed a bland grain-based diet, their bone chemistry will be altered too. Compared with more ancient cluckers, the fossilised bones of modern chickens will contain different ratios of dietary isotopes, and if future geologists were to probe further, and analyse the DNA inside these bones, they would find that their genetic makeup has also changed. Aside from the reduced levels of genetic diversity that potentially leave today's chickens more vulnerable to disease, we have selected for DNA sequences that influence the birds' response to light and its metabolism. As a result, our modern broiler chickens can reproduce year-round and are always hungry, so they eat and grow more quickly.

Multiple lines of evidence – biology, genetics and bone chemistry – will alert future geologists to the animals' significance. They will recognise a bird that evolved quickly; that was absent in the Holocene but everywhere in the Anthropocene. Billions of chicken bones end up in burial tips and landfill sites every day, where the anaerobic conditions mean that often, the remains tend to mummify rather than rot away. In the future, their fossils are likely to be commonplace, so perhaps it will be the humble chicken that comes to define the Anthropocene, and our modern broiler birds will provide that elusive golden spike. In 2008, South Koreans were forced to euthanise 10 million poultry after repeated outbreaks of avian influenza ripped through the country. Their remains now lie in mass burial sites where it's entirely possible they are sliding slowly towards fossilisation. Maybe one of these will be the location of the Anthropocene's golden spike, and when future geologists realise what they have found, maybe they won't call it the Anthropocene after all. Maybe they will call it the Gallicene Epoch – the Age of the Chicken.

Who Gives a Cluck?

All domestic farm animals are products of this new human-influenced epoch. They are creatures of our making and they only exist in their current form because humans have made them that way. This is a clear-cut example of human-mediated evolution. Domestication fuelled the rise of civilisation and nourished our forebears as our numbers grew, but it also came at a price.

Today, humans benefit greatly from the domestic animals we have created. They provide us with food and secondary products like leather and wool, and they line the pockets of those companies that mass-produce them. Three companies, for example, supply 90 per cent of the world's broiler chicks and the poultry industry is now worth billions of dollars. Humans have won and chickens have lost; but if only it were that simple. In our rush to mass-produce cheap meat, we have turned a blind eye to the wider consequences of industrial-scale farming. The way we produce our food has become an ecological disaster that is nudging evolution onto a crash course with extinction. The way we eat is killing the planet.

For the vast majority of its 10,000-year history, agriculture was an outdoor pursuit. Animals were kept alfresco where they could graze and browse, and produce fertiliser which then helped to nourish the crops. Chickens foraged for grain and other titbits, while the bigger ruminants delighted in the sun-nourished stems and leaves that they nibbled. Farming and nature existed side by side in a relatively harmonious fashion, but by the 1950s, a perfect storm was reaching its climax.

The storm began in the early 1900s when chemists worked out how to convert atmospheric nitrogen into ammonia using hydrogen and a catalyst. The result was a material that could be used to make both explosives and fertilisers. Then in the Second World War, German scientists discovered how to mass-produce organophosphate nerve agents, which could also be used as pesticides. In the space of 40 short years, agriculture had acquired two new bedfellows in the form of substances that nourished the soil and killed pests. Corn production expanded massively and new varieties of high-yield crops began to emerge thanks to the practice of atomic gardening (see Chapter 3). Now instead of using animals to fertilise the ground, farmers used chemicals and became indebted – literally – to the companies that supplied them. They found themselves shelling out for seeds, fertilisers, pesticides, antibiotics and equipment, all the trappings of industrial agriculture; so while their productivity and their gross income increased, their net income declined. Farming went from being a relatively small, self-contained, sustainable business to an endeavour that relied heavily on the products of industry. The old system of rotational land use, where arable alternated with animal, became a thing of the past, and as monocultures of high-yield crops spread around the globe, farm animals found themselves squeezed out. As the storm reached its peak, the animals were moved indoors to free up space for crops to grow.

Philip Lymbery, wildlife lover and Chief Executive Officer of Compassion in World Farming (CIWF), calls it the first in a series of great disappearing acts. Chickens, ducks, turkeys, pigs, goats, cattle, sheep and fish vanished from the countryside only to reappear inside purpose-built cages, crates, pens, feed-lots and hangars. Factory farming was born. 'Cramming animals into cages and crowded sheds may have seemed like a space-saving idea,' says Philip, 'but it ignores the fact that a vast amount of land is required elsewhere to grow food for them – often in vast crop prairies doused in chemical pesticides and fertilisers.' Land that was previously used for grazing and foraging – converting things that people can't eat, like grass and weeds, into things that people can eat, like milk and meat – was now used to grow animal feed. 'The food system became hijacked by the animal-feed industry,' says Philip. 'Growing animal feed became a massive operation in its own right.'

Instead of letting farm animals find their own food, factory farms now had to provide it for them, 24/7. So they turned to land crops such as corn, soya and palm, and ocean-derived foods such as fishmeal. Across the globe, enormous swathes of land were given over to growing animal food, while huge shoals of fish were plundered from the sea. This triggered the second wave of vanishings. Trees, bushes and hedgerows were replaced with fields full of pesticide-doused crops, which led to the disappearance of wildflowers, insects and farmland birds. In the last 45 years, British farmland birds have more than halved in number, representing an estimated loss of around 44 million individuals.

Elsewhere, pristine forest was and still is being butchered to make way for palm and soya. Philip recalls a plane journey that he took from São Félix do Araguaia – a place the locals call 'the end of the Earth' – in central Brazil. 'We took off over this wonderful rainforest that stretched away. I watched the Paraguay River wiggle away like a giant serpent. As we flew, I looked down and started to notice the forest begin to ebb away; just small patches at first, then larger ones. Then whole chunks were gone. The forest became an island in a sea of soya crop, and then the forest was gone. I was watching the lungs of the earth disappear. I remember being transfixed by this vast landscape with not a single tree or hedge anywhere. What I did see from time to time were teams of large combine harvesters sweeping the landscape like aerobatic teams. It was then that I realised the sheer scale of what was going on.'

In 2018, Brazil overtook the USA to become the world's biggest producer of soya. The South American country grows around 117 million tonnes of the crop every year, much of which is exported to China and the European Union, where it is used to feed industrially reared animals. This is at the direct expense of the country's dazzling wildlife. Home to the Amazon

Left: The pizzly bear is a cross between a polar bear and a grizzly. This is Tips, a pizzly bear from Osnabrück Zoo in Germany. As the world warms and the ranges of the two parent species increasingly overlap, hybrids like this are becoming more common in the wild.

elow: Peppered moths changed colour during the Industrial Revolution hen a well-timed mutation led to the emergence of a sooty-coloured rm. My Great Uncle Rick's peppered moth is the fourth up from the ottom right hand corner.

Above: Meet Pom Pom (left), and Simon Cowell, our two bouffant bantams. Products of selective breeding; they have feathery feet, an extra toe and hairstyles that belong in an eighties pop video.

Left: The red canary is th world's first transgenic animal. This perky little flapper contains the genes of two species; the domestic canary and the wild red siskin.

Above: Spot the kākāpō, the world's rarest, fattest, least-able-to-fly green parrot. It brays like a donkey, wheezes like an asthmatic and booms like the baseline of a house music anthem.

Below: Humans are directly responsible for the genesis of the Italian sparrow. When Asian house sparrows followed early farmers into Europe around 8,000 years ago, they met and mated with the Spanish sparrow, creating this new hybrid species.

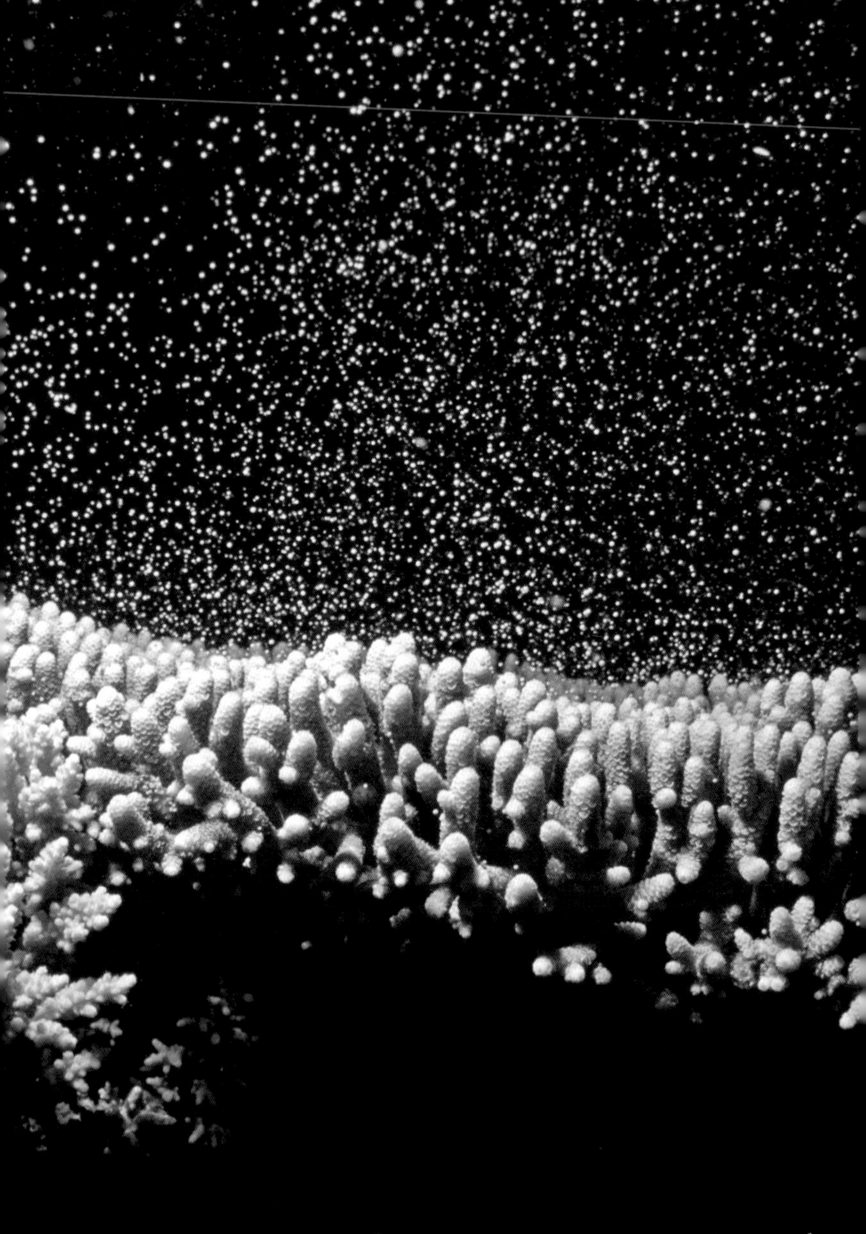

Above: Broadcast coral like *Acropora millepora* spawn synchronously. Each tiny ball contains around 8 eggs and tens of thousands of sperm. These are being used for coral IVF in an attempt to engineer more resilient reefs.

bove: Sea-monkeys aren't monkeys, and they don't live in the sea. hese tiny crustaceans are humble brine shrimp, but they became a global henomenon after an American scientist created a new hybrid variety.

elow: Fancy a transgenic tropical fish? GloFish® contain added genes om coral and jellyfish. This is the Galactic Purple® Shark.

Above: At the Knepp Estate, English longhorn cattle act as a proxy for the extinct aurochs.

Left: Male purple emperor are gutsy creatures. Adults fortify themselves with fermenting tree sap, and males spend much of their time fighting in drunken duels.

Left: Tamworth pigs kick back and relax in a ditch at the Knepp Estate. By disturbing the earth, the pigs create the conditions for the purple emperor's food plant to grow. Now this rare British butterfly i making a comeback.

bove: This little foal is Paris Texas, the first cloned horse in North
merica. He stands next to Greta, the mare who gave birth to him. He
as created over ten years ago by Katrin Hinrichs and colleagues from
exas A&M University, but now he lives in Denmark.

elow: The world's first flock of CRISPR-edited sheep. Scientists used
RISPR to beef up a breed best known for its superfine wool. The result
an animal that can be used for both meat and knitwear.

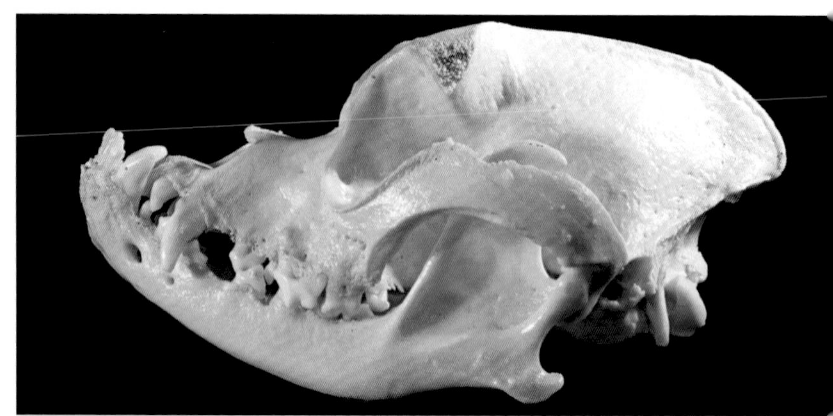

Above: The grimacing skull of a bulldog; selectively bred for strength, agility and the ability to pin a tethered bull to the ground. The undershot jaw helped it grip it opponent's face, whilst the nose is set back so the dog can breathe as it holds on.

Left: Meet Higgs, my genetically modified wolf. He sleeps on the bed, eats cheese and barks at bin bags.

Below: At Heythrop Zoo, the animals are trained via positive reinforcement. Here, Glacier the silver fox (right) has trained me to feed her a piece of chicken when she looks at me adoringly.

rainforest and the huge savanna-woodland region known as the Cerrado, Brazil is the most biodiverse country on Earth. One-tenth of the world's species live here, and many of them are found nowhere else. Now, iconic species like the jaguar, giant anteater and giant armadillo are slipping towards extinction because their homeland is being eroded to produce animal feed.

It's a similar story in Indonesia and Malaysia. In Sumatra, more than half of the island's lowland forests have been burned in order to make way for palm plantations. The bogeyman of household shopping, palm oil is found in around half of all packaged supermarket products, including lipsticks, shampoo and ice cream. Production has more than doubled since the turn of the millennium, but while the public is increasingly aware of its presence in packaged items, few are aware of the role that palm plays in factory farming. While the flesh of the fruit is used to make palm oil, the protein-rich kernel is ground down to make palm-kernel meal, which is then used to supplement the diets of factory-farmed cattle, pigs and poultry living thousands of miles away. Meanwhile wildlife is struggling. Just as in Brazil, these island refuges are home to many unique species that are simply irreplaceable. Javan rhinos, Sumatran elephants, tigers and the 'old man of the forest', the orangutan, are also facing an uncertain future, because their homeland is being eroded to produce animal feed.

On South Africa's Cape Peninsula, endangered African penguins feed almost exclusively on anchovies, pilchards and red-eye. In recent years, however, their favoured foodstuffs have been in short supply as commercial fisheries hoover hundreds of thousands of tonnes of the small pelagic fish out of the ocean. What was once the region's most numerous bird is now in serious danger. There are just 50,000 African penguins left in the wild and some fear that in 15 years' time, they could all be gone. 'And it's not just the penguins,' says Philip, who has witnessed the situation first hand. 'Everything else that relies on the small, pelagic fish is also affected – all the way from hake and yellowtail, to sharks and tuna, to seals, dolphins and whales.' One way or another, all these marine dwellers depend on the little guys for their survival, and now all are feeling the pressure. But the little fish are not being caught for human consumption. Most of the catch is turned into fishmeal and exported to feed factory-farmed animals.

Today, two-thirds of the world's 70 billion farm animals are reared on factory farms where they munch their way through a third of the world's cereal harvest, 90 per cent of its soya meal and up to 30 per cent of its global fish catch. Crops and fish that could have been eaten by people are disappearing to feed domestic farm animals. The way we produce our food has become a profoundly wasteful and inefficient process. It's wasteful because much of the food we produce is never eaten. Every year, consumers in rich countries discard over 200 million tonnes of perfectly edible food. It's inefficient because much of the nutritional value hidden inside crops like maize, palm and soya is lost as it is converted, via animals, to meat, milk and eggs. According to the UN Food and Agriculture Organisation, when livestock are raised intensively, they consume nutrients that could have been eaten directly by humans and use them to produce smaller amounts of energy and protein. For every 100 calories of grain fed to animals, we receive only about 40 new calories of milk, or 22 calories of eggs, or 12 calories of chicken, or a measly 3 calories of beef. Energy is being wasted.

Compassion in World Farming estimates that if the world's grain-fed animals were all returned to pasture, and the cereals and soya previously used to feed them went to people instead, there would be enough food to feed an extra 4 billion people. If the fish being used to feed farm animals were instead fed directly to people, there would be enough to feed a billion people. It would help to ease the pressure on the natural world too.

When Philip first joined Compassion in World Farming 30 years ago, factory farming was predominantly a welfare issue. The cramped and unnatural conditions experienced by industrially reared animals sparked condemnation from people who believed that animals deserved better. Today, however, factory farming is no longer just an animal welfare issue. We're all aware that deforestation, pollution and poaching are leading to the extinction of our wildlife, but it's now becoming increasingly obvious that factory farming is also playing a big role. Agriculture is one of the biggest drivers of biodiversity loss in the world, second only to the non-sustainable, over-exploitative harvesting of wild species. Two-thirds of species listed as threatened or near-threatened on the IUCN Red List* are at risk because of

---

* The International Union for the Conservation of Nature (IUCN) has the unenviable task of managing the world's Red List, a compilation of species categorised according to how common or rare they are. The future is not rosy for species that are listed as 'threatened' or 'near threatened'.

the way we produce our food, and in 2019, a special commission from the prestigious *Lancet* medical journal warned that our current food system is putting Earth's ecosystems at risk. It's insane, especially when you realise that we don't need intensive agriculture in order to feed the planet. 'Feeding human food to factory-farmed animals is perhaps the greatest disappearing act of all,' he says.

'The other problem,' says Philip, 'is that industrial agriculture is undermining the ability of future generations to produce food at all.' Raising livestock for meat, eggs and milk generates 14.5 per cent of global greenhouse emissions. This is more than all the world's planes, trains and cars put together. Factory farming is fuelling climate change, which in turn is leading to biodiversity loss. 'We are seeing ecosystems collapse. We are seeing run-off from chemical-soaked fields polluting waterways and creating dead zones that are growing in our oceans. We are seeing pollinators that are essential for a third of our food crops in steep decline. If current trends continue, commercial fisheries could run dry by 2048. The UN has warned that if we continue as we are, the world's soils will have effectively gone within sixty years. No soils, no food. It's as simple as that. We are looking at Farmageddon.'

A Glittery Wilderness

Let's recap. Ten thousand years ago, we began to sculpt the genomes of the animals that we would later incarcerate in factory farms. Among other things, we fashioned cattle from aurochs and chickens from red jungle fowl. These animals were selectively bred for the ability to grow large quickly, often at the expense of their own wellbeing. For a long time, we shared the earth amicably with these creatures of our making, but in the last 70 years, things changed dramatically. In days gone by, Old MacDonald had a farm … with a cluck cluck here and a moo moo there. All the animals lived outdoors where they got to graze and forage and fertilise the ground. Now the vast majority of farm animals are hidden away inside factory farms and agriculture has expanded to take up nearly half of our world's useable land surface. Farming has changed the face of the Earth, and farm animals are now the most numerous large, land-living group of animals on the planet. The rise of domestication led to the rise of agriculture, which in time helped fuel the demise of the natural world. This is an epic evolutionary story of our own making. In 100 years' time, the world's wild megafauna could all be gone and the largest land-living mammal will probably be the domestic cow. Meanwhile chickens will probably remain the world's most numerous bird, and billions of them will live out their short lives within the confines that we dictate. So the questions are twofold: where should our 70 billion farm animals live, and what should they eat?

When I first met Philip, he looked like a fish out of water. He was the only suited and booted man at a festival where many people wear little more than shorts and body glitter. We were at Wilderness, a family-friendly music festival in the heart of the UK's leafy Cotswolds. It was a roasting summer's day as the audience made their way into a sweaty marquee and squashed up next to one another on hay bales.

A composed and erudite spokesman, Philip explained to the crowd how wildlife is paying the price for factory farming. I like Philip. He doesn't preach and he doesn't get angry. He simply explained the chain of cause and effect that has led us to where we are. The stories he tells are gathered first-hand, via his travels around the globe and the investigations he has made into the agricultural industry. The statistics that he quotes are underpinned by rigorous, peer-reviewed academic studies. The conclusion is obvious. We need to change the way that we produce our food. 'For a long time, the environment seemed able to absorb the heavy knocks associated with this type of farming,' Philip told the attentive audience, 'but I think that we are now reaching a tipping point. The Earth's ability to take punishment – be it pollution or climate change or the draining of natural resources – is reaching its limit. Agriculture has a big part to play.'

In this Age of the Chicken, perhaps it's time to let our animals live outdoors again, where they have sun on their backs, space to potter and natural food to eat. There is space for them. The amount of land currently being used to grow animal feed covers an area equivalent to the entire land surface of the European Union. And when we put farm animals back on the land, amazing things start to happen, Philip told the festival-goers. Farm animals become able to express their natural behaviours. They can stretch their wings or roll in the mud. It's good for their wellbeing. When put to pasture in a mixed rotational system, they help to improve impoverished soils and the landscape starts to regenerate. Wildflowers grow and wildlife returns. Critically, this takes the pressure off the forests – near and far – that are being cut down in order to mass-produce animal feed. This helps to preserve the homes of iconic wildlife species like jaguars, orangutans and Sumatran elephants.

A snoozing child stirred in her mother's arms. It was hot and sticky in the festival tent but the adults stayed focused. They leaned forwards on their hay bales as if to drink in Philip's words. 'So what can we do?' he asked rhetorically. 'I believe we are the last generation that can leave a planet worth having as a legacy for our children. The great news is that we can all help to bring this regenerative food system about through the food choices we make three times a day. We can eat more plants, and less and better meat from pasture-fed, free-range and organic systems. This brings a life worth living to farm animals, helps save iconic wildlife, and preserves the sustainability and quality of our food for us today, and for our children tomorrow.'

Small changes can make a big difference. This is something that I truly believe. An enthusiastic round of applause erupted as bums shuffled on bales and sticky bodies began to make their way out of the crowded tent. Outside I noticed the long queue for a vegan food stall curling round the side of the tent. Philip is right, I thought to myself, good food doesn't have to cost the earth. But *six pounds* for a bean burger…?

CHAPTER SEVEN

Sea-Monkeys and Pizzly Bears

In the mid-1960s, an eye-catching advert began to appear in American comic books. It featured a nuclear family of grinning aliens with pronged heads and pot bellies. The humanoid figures had scaly chests, webbed digits and long swishy tails. Save for the ribbon in Mother Alien's crown they were also completely naked, but to spare the readers' blushes, their private parts were obscured by carefully placed flippers. They posed in front of a fairy-tale castle in an underwater world and promised a 'bowlful of happiness' for anyone prepared to part with US$1.25 (£1) and an additional 50 cents (40p) for shipping. The accompanying text read:

> Enter the WONDERFUL WORLD of *AMAZING* LIVE SEA-MONKEYS

The advert was the brainchild of a colourful character called Harold von Braunhut, who had his fingers in many pies. He raced motorbikes as 'The Green Hornet', and managed the career of a showman whose act involved belly-flopping 12 metres into a paddling pool. A serial entrepreneur, von Braunhut created a string of popular children's toys, including X-Ray Specs – glasses that promised to help you see through clothing and skin – and the Invisible Goldfish – an imaginary pet you were guaranteed never to see. In a time where families were bombarded with TV adverts for the likes of Barbie and hula hoops, von Braunhut decided to bypass the parents and sell directly to the kids. So he hired Joe Orlando, who would later become Vice President of DC comics, to do the artwork and snapped up millions of pages of advertising in comic books like *The Amazing Spider-Man* and *Casper the Friendly Ghost*.

Von Braunhut's colourful style spoke directly to a generation that was unfettered by today's modern trappings. In a world without mobile phones or the diversions of the Internet, what child could fail to be lured by the promise of sea-monkeys: 'so eager to please, they can even be trained' and 'obey your commands like a pack of friendly trained seals'. Kids started to cough up their hard-earned pocket money and an advertising phenomenon was born.

Of course, sea-monkeys aren't actually monkeys, nor do they live in the sea. The tiny creatures are brine shrimp, a diminutive crustacean that can be found living in salty inland lakes. Their segmented bodies are flanked by eleven pairs of 'legs', and their long tails reminded von Braunhut of monkeys – hence the name. Fascinating creatures; they breathe through their legs, have three eyes, and can live in an egg-like state of suspended animation for years at a time. It's a phenomenon known as 'cryptobiosis' and Harold sought to capitalise on it when he promised his customers 'Instant Pets – Just Add Water'.

The product came with a manual and two sachets. Packet One contained a 'water purifier', while Packet

Two, which was to be added 24 hours later, contained 'instant live eggs'. When the latter was added – boom – instant life! The sea-monkeys miraculously appeared as if from nowhere and started cavorting around in the water.

The only problem was that in its early iterations, the product was spectacularly underwhelming. Very few of the eggs ever hatched and when they did, the brine shrimp often died quickly without so much as a sea-monkey sayonara. Children did not take the death of their pets well. Tears and tantrums followed, prompting Harold to enlist the help of a scientist called Anthony D'Agostino from the New York Ocean Science (NYOS) Laboratory in Montauk.

D'Agostino was a marine biologist specialising in brine shrimp. More than anyone, he knew what the critters like to eat and how to keep them in captivity. He had been studying different strains of brine shrimp, *Artemia salina*, sourced from various parts of the world including the Great Salt Lake of Utah, USA and the Comacchio salt ponds near Venice, Italy. Working together, D'Agostino and von Braunhut began hybridising these various strains until they managed to create a hardy brine shrimp that could survive in its egg-like state for prolonged periods of time and that thrived after reanimation. Never a man to shy away from bold statements, von Braunhut declared that they had created a new species. *Artemia salina* gave way to *Artemia NYOS*, named after the laboratory of its birth.

After that, sea-monkeys went from strength to strength. The new, improved brine shrimp could survive longer in the post as eggs, and lasted long enough after hatching to satisfy the kids who bought them. In 1972, the miracle of 'instant life' was revealed in a patent which described the true contents of the enigmatic sachets. Packet One did indeed contain a water purifier, but it also contained dormant brine shrimp eggs, which quickly hatched on contact with water. Packet Two contained extra eggs, but was also loaded with a blue dye, which then settled on the transparent day-old sea-monkeys that had already hatched and made them instantly visible. There was no 'instant life', rather a clever bit of marketing and the illusion of instant life.

Children didn't seem to mind that the life wasn't really instant, nor that their new pets bore absolutely no resemblance to monkeys, friendly trained seals or the aliens from the adverts. Parents seemed to tolerate the misleading adverts* and soon sea-monkeys had found their way into thousands of American homes. Never one to miss a trick, Harold started marketing add-ons, like a 'Banana Treat' desert and a sea-monkey aphrodisiac he called 'Cupid's Arrow'. He built an unlikely empire and a cultural phenomenon from a humble crustacean more commonly used as fish food. To this day, sea-monkeys have been featured in *The Simpsons* and *South Park*, and even though von Braunhut has passed away, the business is still worth millions of dollars.

In the grand scheme of evolution, sea-monkeys may seem like an irrelevance; more of a gimmick than an

* And the fact that Harold has been linked to an anti-Semitic, neo-Nazi, white supremacist hate group called the Aryan Nations. He pledged them the profits from another invention of his: a weapon called the Kiyoga spring whip, which he also marketed in comic books.

evolutionary milestone. Yet here is a species that has been deliberately altered, not for agriculture or medical research, but so that it pleases children. 'To the best of my knowledge it's the only living creature that's ever been specifically bred so that its lifecycle matches the lifecycle of a toy,' says Richard Pell from Pittsburgh's Center for PostNatural History, which features a rather spectacular sea-monkey display. Yes, von Braunhut had a talent for exaggeration, but he was also an imaginative inventor and a rigorous scientist. 'On the one hand, Harold was operating as a sort of twenty-first-century P. T. Barnum, but at the same time the invention and experimentation that went into making sea-monkeys work was a lot more like a Thomas Edison,' says Richard. It's precisely because the story seems so frivolous, that it stands out. Whether or not *Artemia NYOS* qualifies as a new species is a matter for debate, but it is undeniably different from the classic *Artemia salina*. It deserves our attention because, without von Braunhut and his bonkers advertising, this unique little creature simply wouldn't exist. If it's flushed down the toilet or poured into a pond, it doesn't survive, meaning that here is a lifeform, engineered by humans for children, that simply does not exist in the natural world. What a legacy for von Braunhut and D'Agostino to leave behind.

Pizzlies, Shetbras and Wholpins

When individuals from two different strains, breeds, varieties or species mate and then have offspring, the result is known as a hybrid. Genomic testing has revealed that hybridisation is commonplace in the natural world, where 10 per cent of animal and 25 per cent of plant species are involved in hybridisation. This includes 75 per cent of British duck species, and 12 per cent of European mammals and butterflies. Many corals hybridise too. Hybridisation is part of the natural story of life on Earth, but this process is also influenced by humans. *Artemia salina* is a hybrid created at the hands of man. Similarly, when Hans Duncker tried to create the enigmatic red canary, he started by making hybrids of yellow canaries and red siskins (see Chapter 3). Sometimes breeders and zoo keepers deliberately create hybrids as a source of novelty, and sometimes hybrids pop up by accident when species housed in close proximity get closer still. Many readers will have heard of mules and hinnies, the star-crossed offspring of donkeys and horses; not to mention ligers and tigons, which are products of lions and tigers. Now, as human-induced climate change alters the natural range of living creatures, and as we manoeuvre species around the globe, we are creating novel opportunities for species to meet and mix. The result is that unusual wild hybrids are starting to rear their heads, all thanks to the actions of humans.

Wig-wearing taxonomist Carl Linnaeus was one of the first to describe hybrids, back in the eighteenth century, when he noted how different species can interbreed to create forms that look like new species. The doyen of evolutionary theory, Charles Darwin, was himself fascinated by these inter-species hybrids. He devoted an entire chewy chapter of *Origin of Species* to the subject, where he cogitated over their inconsistencies. He points out that species are fluid and that hybridisation does occur, then goes on to list all the ways that hybridisation is confusing and contradictory. Some species hybridise, others do not. Sometimes physical similarity can be used as a guide as to whether two species will hybridise, sometimes it can't. Sometimes the males of species A hybridise more successfully with the females of species B. Sometimes it's the other way around. Sometimes the offspring of hybrids are fertile. Sometimes, you guessed it, they're not.

Today, as the prevalence and importance of hybrid species becomes increasingly apparent, these conundrums are starting to be resolved. We know, for example, that the way the DNA is bundled up inside the cell's nucleus has an effect. If parent species have different numbers of chromosomes, then successful hybridisation is less likely and any resulting offspring are likely to be sterile. So lions and tigresses, which have 38 chromosomes each, can have fertile liger offspring, but horses and donkeys, with 64 and 62 chromosomes respectively, tend to produce offspring that are infertile. In the plant world, whole sets of chromosomes can become duplicated in a process known as polyploidy. Several commercial fruits, such as loganberry and grapefruit, are polyploid hybrids, as are garden herbs such as peppermint, and trees like the London plane, which is a cross between the American sycamore and the oriental plane. Not to be outdone, common wheat is a hybrid containing not two, but three complete sets of chromosomes inherited from three different wild grass parents.

Compared with animals, plant hybrids are easy to make, so it's no surprise that humans have created an enormous number of hybrid varieties. Some have been crafted by gardeners as they lovingly toiled in their greenhouses, others by farmers and food scientists.

Peanut plants are a hybrid of two different parent species, as are three new *Brassica* species: rapeseed, Indian mustard and Ethiopian mustard. According to the United Nations Food and Agriculture Organisation, 6 of the world's 40 most important food crops are human-created hybrids.

When humans create hybrids, we aim to produce something that is superior to the parent species. Our hybrid corn, for example, tends to have a higher yield, is better at withstanding drought and disease, and is more nutritious than its non-hybrid ancestors. Meanwhile, mules are said to be more intelligent than donkeys and hardier than horses. Darwin was a big mule fan. He wrote of them: 'The mule always appears to me a most surprising animal. That a hybrid should possess more reason, memory, obstinacy, social affection, powers of muscular endurance, and length of life, than either of its parents, seems to indicate that art here has outdone nature.' Dependable and hard-working, the mule's physique and temperament have made it a favoured beast of burden and companion to soldiers at war.

Before we continue, let me introduce you to the portmanteau tradition of naming animal hybrids. Portmanteau words are themselves hybrids of the words they are derived from. Classic examples are smog (a mix of smoke and fog) and brunch (a mix of breakfast and lunch), but current examples include cellfish (someone who talks on their phone in the quiet carriage), askhole (someone who repeatedly asks pointless questions) and nerdjacking (filling a conversation with unnecessary detail about science*).

---

* Guilty as charged.

Hybrid animals are not always given portmanteau names, but when they are, the results are delightful. The convention is that the father provides the front half of the word, while the mother brings up the rear. So a liger is the offspring of a lion and a tigress, while a tiger and a lioness may give birth to a tigon. When a male coyote liaises with a female grey wolf, the result is a coywolf, but if the provenance is reversed, a woyote pops out. There are also zonkeys (zebra meets donkey), pumapards (puma meets leopard) and gruffalos (American buffalo meets bedtime story).

In captive settings, humans can deliberately engineer hybrids if the parent species are closely related and if Cupid's arrow hits the right spot. Mules, or 'dorses' as they should really be known, are the oldest human-created hybrid, first bred by people living in Paphlagonia (now northern Turkey) over 2,000 years ago. More recently, in the nineteenth century, farmers fashioned beefalo to maximise the hardiness of the buffalo with the milk-making abilities of the domestic cow, while camas were created in the nineties when scientists artificially inseminated llamas with camel semen.

When hybrids appear in public zoos and private collections, it can be a mixed bag. Worldwide, around 40 public zoos display ligers, with many more kept by private owners behind closed doors. Separated by about 7 million years of evolution, lions and tigers seem to be on the very cusp of genetic compatibility. According to the animal welfare charity PETA, ligers can experience health problems such as cancer, arthritis and organ failure, and because the newborns are so big, labour can be a dangerous process. When ligers do survive, there's no denying that they are impressive beasts. Subtly striped, adult ligers can grow to about 1.5 metres (5ft) tall. They can weigh up to 450kg (990lb), roughly the same as both their parents combined, making them the largest cat on Earth. The world's largest liger, a male called Hercules, is so big that he even has his own website, but don't go there … not unless you enjoy seeing photos of rare animals being taken for a walk on a chain. Reputable zoos do not risk the health of endangered species by breeding them together to create pointless genetic novelties, yet there are breeders out there who deliberately create endangered hybrids because it's profitable. Big cat hybrids do not occur naturally. Lions and tigers don't generally overlap in the wild and in the one rare place where they do – India's Gir Forest – there are no ligers. The two species give each other a wide berth. For as long as there are healthy lions and tigers still alive that are able to mate with members of their own species, there is no conservation value to breeding ligers.

That said, hybrids do sometimes crop up unintentionally in zoos, farms and private collections when seemingly disinterested species are housed in close proximity. When keepers at Hawaii's Sea Life Park put a false killer whale and an Atlantic bottlenose dolphin into the same enclosure, they didn't think for a moment that anything untoward would happen. Although false killer whales are members of the dolphin family, the two cetaceans are separate species that belong to different genera.* They're also very different sizes. An adult false

* A genus (plural genera) is the taxonomic rank that is one up from a species and one down from a family.

killer whale can weigh twice as much as an adult Atlantic bottlenose dolphin, but the two residents showed that sometimes, size really doesn't matter. Their healthy hybrid daughter, Kekaimalu (meaning 'from the peaceful ocean'), was born in 1985. Meet the wholpin.

In 2004, a new arrival gave South African farmer Tom Beckett the surprise of his life when a little Shetland pony called Linda gave birth to a foal with black and white stripes. Yet again, size proved no barrier – literally – when a male zebra called Jonny leaped over the fence that separated him from Linda, and struck up an affair with the diminutive equine. Less than a year later, Nikita the Shetbra was born.

The same year, twin bear cubs were born in Germany's Osnabrück Zoo. Mum was a grizzly bear* called Susi, while Dad was a polar bear called Elvis.† The two had lived together in the same enclosure for 24 years without so much as a whiff of romance, but in the end they couldn't help falling in love. The result was a pair of pizzly bear cubs called Tips and Taps.

The non-identical, boy-girl cubs were a total delight. Taps had toffee-coloured fur with silver ears, while Tips had silver-coloured fur with darker patches around her eyes, legs and paws. Both had the slender neck of a polar bear, but the shoulder humps of a grizzly. They were smaller than their dad, but larger than their mum, and their feet were somewhere between the flat paddles of a polar bear and the clumpy, clawed stompers of a grizzly.

---

* Scientists don't like the name 'grizzly' bear. They prefer North American brown bear.

† So-called because he was born in Memphis Zoo.

They lived in a large enclosure with pools, a waterfall and a pack of silver foxes, but no one thought for a moment that oddities like this could ever occur in the wild. After all, polar bears are creatures of the frozen far north, while grizzly bears stop short of the Arctic Circle. They may be neighbours, but they have different lifestyles and feeding habits that restrict their range.

Then, in 2006, American hunter Jim Martel shot an odd-looking bear during a licensed polar bear hunt in Canada's Northwest Territories. Although it looked like a polar bear from a distance, up close and personal, Martel and his Inuit guide noticed that the animal had dark, furry smudges around its eyes and muzzle. Its claws were unusually long and the bear had oddly humped shoulders – features more commonly seen in grizzly bears. A biopsy sample was removed and sent for genetic analysis, and when the results came back they confirmed what the hunters had suspected. Just like Tips and Taps, the animal was a hybrid.

Although there had been sightings of unusual-looking polar bears before, this was the first definitive proof that pizzly bears exist in the wild, but how they came to be remains a mystery. Polar bears court and mate on sea ice during the months of April and May, when grizzlies are meant to be far away on terra firma. Also, male and female polar bears need to spend several days in close proximity for the female to ovulate and for mating to occur. This means that the pizzly's father (the polar bear) must have spent time 'courting' the pizzly's mother (the grizzly). He may even have fought off other rivals along the way. Clearly this was no simple one-night stand.

Scientists currently have no idea how common wild pizzly bears are, but they're not the only unusual hybrids to be spotted in the Arctic. In the late eighties, hunters in Greenland reported an odd-looking whale that seemed to be half narwhal, half beluga.* In 2009, a scientist photographed a large cetacean that seemed to be part right whale, part bowhead whale, in the Bering Sea.† Dall's porpoises are known to mate with harbour porpoises off the coast of British Columbia, and seal hybrids have been identified in museum specimens and in the wild. Writing in *Nature*, scientists pointed out that these animals are probably the tip of the Arctic hybrid iceberg. Brendan Kelly, Andrew Whiteley and David Tallmon counted at least 34 possible hybridisations that have already occurred between various Arctic and near-Arctic marine mammals, and there could be many more.

Lethal Pizzle?

So what to make of these mixed-up matings? As we continue to warm our world, a continent-sized barrier to reproduction is beginning to thaw. The Arctic is becoming a global melting pot and hybrids are set to become more common. Humans are unequivocally responsible for this, so you could say that these new Anthropocene hybrids are our doing. They're not unnatural, because hybridisation is commonplace in the natural world, but they are 'post-natural' because humans toppled the first domino in a chain of cause and effect

* A narluga? Bewhal?

† A righthead whale?

that led to their appearance. Now only time will tell what their fate will be.

Kelly and his colleagues are concerned that increasing levels of hybridisation will threaten biodiversity in the Arctic. It's feared, for example, that there are just a few hundred North Pacific right whales left. The gentle giants are slow breeding so every calf counts, but if they mate with one of the more numerous bowhead whales they lose a crucial opportunity to expand their numbers. Similarly, hooded seals could become threatened by hybridisation with harp seals, and polar bears could become threatened by hybridisation with grizzlies.

It's a worry echoed by many conservationists working south of the Arctic Circle. In Scotland, the rare Scottish wildcat is threatened by hybridisation with domestic cats, while in south-east Asia, the precursor of the modern chicken, the red jungle fowl, is threatened by hybridisation with its domestic descendants. Finely tuned by millions of years of evolution, a species' DNA endows it with many unique traits that are vital to survival. Modern chickens, for example, succumb easily to disease, but wild jungle fowl are almost entirely resistant to highly virulent avian flu strains that currently affect poultry flocks in Europe, Asia and North America. As the jungle fowl's genome becomes eroded through hybridisation, the fear is that the DNA underpinning these characteristics will become lost, and that one day, when we need it most – say, to breed new strains of disease-resistant chicken – the resource will be gone.

The flip side is that sometimes when populations are really small and desperately inbred, hybridisation can offer a genetic lifeline. Sometimes it's better to save 50 per cent of a rare species' DNA, than to have its DNA disappear altogether when the species goes extinct. This is something the scientists trying to save the northern white rhino are having to think about. A gentle giant of the central African plains, the northern white rhino has been driven to the brink of extinction through decades of civil war and illegal poaching. As I write these words, there are currently just two individuals left: a mother and her daughter – Najin and Fatu – who live in a wildlife reserve in Kenya. Both have health problems and neither can reproduce naturally, so Thomas Hildebrandt and colleagues at the Leibniz Zoo and Wildlife Research in Berlin are trying assisted reproduction to save the species.*

In 2018, the team reported they had created test-tube embryos using frozen sperm collected from one of the last male northern white rhinos before he died, and eggs collected from a closely related species called the southern white rhino. Two little hybrid embryos developed in culture for more than a week, before the team froze them away to await the next, as yet unrehearsed stage of the process: implanting them into the womb of a surrogate southern white rhino. It's an outstanding

---

* Researchers continue to bicker about whether the northern white rhino and its close relative the southern white rhino are separate species or separate subspecies. A subspecies is a poorly defined term that tends to describe related animals that are genetically similar but geographically separate. Scientists recently found that the two varieties are as genetically different from one another as they are from the black rhino, which is definitely a separate species. With this in mind, I refer to the northern white rhino as a species.

achievement. Half of each embryo's DNA comes from a species that is functionally extinct. So you could say this is diluting the northern white rhino's precious genetic code, or you could say it's preserving a significant fraction of the species' genome inside a potentially viable embryo. Along with Hildebrandt and his colleagues, I plump for the latter. The strategy is a way of rescuing valuable genes from a species that is all but gone. In some instances, maybe it's better to have a hybrid than it is to have nothing at all.

Grizzle-Pizzle

Life is tough for polar bears. The far north is warming about twice as fast as the global average, because the melting ice is less able to reflect the sun's warming rays. Polar bears spend most of their time on and around the sea ice, so as the far north thaws, their natural habitat is literally vanishing from beneath their feet. Some scientists estimate that in a few decades there will be no ice left during the Arctic summer. Polar bears are being forced to venture further south, while grizzlies are expanding their territory northwards. Somewhere in the middle, the two species are meeting. They are already being glimpsed together. Hunters have spotted polar bears and grizzlies hanging out in the Hudson Bay area around a freshly killed whale carcass. Human-induced climate change is altering the range of animals, providing new opportunities for normally isolated species to interbreed. Could this be the end of the polar bear as we know it, or the beginning of a new phase in its evolution?

In 2010, another pizzly was shot in the Canadian Arctic. Once again, genetic tests were performed and once again, they confirmed that the animal was a hybrid, only this time they found that it was not a first-generation pizzly. This creature was a second-generation cross. Its father was a grizzly and its mother was a pizzly, making the animal a 'grizzle-pizzle'. To that point, no one knew if pizzly bears were even fertile, let alone breeding in the Arctic. Nor did they know if pizzly bears were able to bring up their young.

In the wild, grizzly bear mums and polar bear mums form strong bonds with their purebred offspring. They stay together for up to 30 months, while the mothers provide nourishment and vital life lessons, but sometimes parents can reject their hybrid offspring. Tips and Taps were snubbed by their grizzly bear mother in Osnabrück Zoo, and in the end, they were reared partly by their aunt – a bear called Ossi – and partly by people. Bear cubs depend on the kindness of elders. So to survive in the Arctic, the little grizzle-pizzle cub must have been looked after. This suggests that some pizzlies at least have the parenting skills required to raise a family, and hints that a pizzly dynasty of some sort is not entirely out of the question.

Hybridisation can lead to various possible outcomes. Sometimes, the hybrid breeds back with one of the parent species. Rather than establishing a new lineage that is a 50:50 DNA blend, as generations unravel, the genome of one species 'wins out' and becomes peppered with little bits of DNA from the other species. Natural selection prunes and sculpts the genome, leading to the creation of hybrids that are mostly one thing and a little bit another.

You may well be the best example of this. If you are of European or Asian origin, then you are approximately

97.3 per cent *Homo sapiens* and 2.7 per cent *Homo neanderthalensis*. Around 40,000 years ago, after leaving Africa and meandering into Europe, your modern human ancestors met and mated with Neanderthals. We may never know how consensual or otherwise these liaisons were, but they resulted in fertile hybrid offspring, who then grew up and had children of their own. In the tens of millennia that followed, Neanderthals went extinct and modern humans went on to conquer the world, but the genes of Neanderthals live on in us today. I recently took a genetic test and was thrilled to find out that I contain slightly more Neanderthal than most. Of the 3.2 billion pairs of chemical letters that make up my personal genome, around 96 million come from our Neanderthal brethren. This makes me 3 per cent Neanderthal and I think it may explain a lot about me: the hairy toes, the love of bonfires and the inexplicable urge I sometimes feel to club annoying people around the head* … but I digress.

In the south-eastern US, red wolves are around 80 per cent coyote and 20 per cent wolf. South American *Heliconius* butterflies have acquired their brilliant colour schemes by breeding with closely related species, while on my home turf in England, the wild-growing rhododendron that I pass on my dog walks contains a smattering of genes from two North American

---

* Apologies to my Neanderthal ancestors. This is a cheap shot. Overwhelming amounts of evidence suggest that Neanderthals were not the brutish thugs of common misconception. They buried their dead, nurtured their disabled and had a complex culture that was probably not so different from ours.

rhododendron species. The organisms are all hybrids, but the majority of their DNA comes from one parent species. When scientists studied the genetic makeup of grizzly bears, a similar picture emerged. Instead of being pure grizzly, the animals carry between 3 and 8 per cent polar bear DNA. Just like so many seemingly 'pure' species, on closer inspection, grizzlies are a genetic blend of more than one species. This tells us a couple of things.

The first is that hybridisation is nothing new. To arrive at this situation, grizzly and polar bears must have been interbreeding on and off for a very long time. Beth Shapiro's team from the University of California, Santa Cruz, estimate that the two species have interbred sporadically for at least the last 40,000 years, and probably for much longer. In the time since polar bears and grizzlies evolved, sometime in the last 500,000 years, the Arctic has gone through various intermittent warm spells. As the ice melted and their ranges overlapped, the two species probably came into contact and had offspring together. Then, when it became colder and their ranges diverged, the animals went their separate ways and returned to breeding with their own kind.

The second thing is that, in the long run, there could be some evolutionary value to these matings. When scientists scrutinised the human genome more closely, they found that the Neanderthal genes we possess are far from random. Natural selection has favoured the retention of genes that are useful to us. When our ancestors arrived in Europe they would have had no immunity to the novel pathogens they found there. Neanderthals, in contrast, who had been living in the region for hundreds of thousands of years, would have been well adapted. Hybridisation allowed us to borrow and then retain certain Neanderthal genes that are involved in the ability to fight off disease, and this in turn has helped us to survive.

Similarly, grizzlies may be benefiting from the polar bear genes that they carry. To date, no one has studied the grizzly bear genome at the level of detail needed to determine the precise nature and function of the polar bear genes that they carry. This will come. In the meantime, what we do know is that while grizzly bears contain polar bear DNA, the reverse is almost never true. Polar bears hardly ever contain grizzly bear genes. This makes sense. Imagine a polar bear with a brown grizzly coat, like Taps from the Osnabrück Zoo. In its natural setting, against a snow-covered backdrop, the animal would stick out like a sore thumb. Seals would see it coming from miles away, so the bear would be more likely to go hungry, less likely to mate and less likely to pass its brown genes on.

In a rapidly changing world, this ability to absorb new genes from other species is a very useful skill to have. The alternative is to wait for new, useful mutations to crop up, but this can be a slow and random business. Hybridisation is a way of generating a much larger amount of new genetic diversity in a much smaller time frame. If polar bears and grizzlies continue to mate, as is expected, then the pizzly dynasty could become a very dilute affair. It may be that polar bears as we know them today vanish, but that their genes live on in their grizzly bear relatives. Or there may be another alternative.

Hopeful Monsters

In the past, evolutionary biologists puzzled over the mechanisms that allow species to make large, sweeping changes to their appearance and biology. Darwin once wrote, 'natural selection acts solely by accumulating slight successive favourable variations, it can produce no great or sudden modification; it can act only by very short steps.' According to the great man, new species evolve slowly and subtly. There was no place in his scheme for swift, elaborate change.

Then along came a German geneticist called Richard Goldschmidt. Goldschmidt did not believe that the small-scale changes envisaged by Darwin were enough to generate bold new features or, indeed, new species. Instead, he proposed that big changes are caused by big mutations. Suppose, for example, that a mutation occurs in a gene that controls the basic layout of the body. Just one error and things could change significantly. *If* the mutation is compatible with life, then the organism that develops could be very different indeed. He called these creatures 'hopeful monsters', in recognition of the fact that sometimes – just sometimes – these mega-mutations might be useful and so retained by natural selection. The hopeful monster would then grow up and become the founding member of a new and unique species.

His critics laughed him out of town. Among other concerns, they wanted to know exactly *where* this hopeful monster would find a willing mate. It doesn't matter how well adapted an organism is to an environment, if it's unable to reproduce, it becomes a genetic cul-de-sac.

Since then, years have passed, and while the Darwinian view still holds sway, scientists now appreciate that there are multiple routes to speciation. It can happen slowly as mutations crop up and are selected for across geological time spans, but it can also happen rapidly. To some degree, Goldschmidt has been vindicated. Although it is rare, hopeful monsters have been glimpsed.* Key mutations in developmentally relevant genes can drastically alter the way an organism develops, but there is another, often unappreciated path to rapid and significant change: hybridisation.

If two species hybridise, and then forgo mating with either of their parent species, it can lead to the formation of a new species. Suppose that in future, a small group of pizzly bears find themselves in a remote part of the Arctic where there are no polar or grizzly bears. They breed with each other and find they are well suited to the environment they live in. The ice has gone and they live on land, where being brown helps the bears to blend in. They are excellent swimmers, like polar bears, but hibernate in winter, like grizzlies, and somehow this mishmash of hybrid characteristics helps them to thrive. They become geographically and genetically isolated from their parent species, until it gets to a point where even if they did meet, they would either have no interest in mating or would be unable to produce viable young together. There's currently no way of knowing if this will happen, but if it does, then perhaps we are witnessing the very early steps of speciation.

---

* For example, it's thought that the stunning yellow-banded day sphinx moth evolved after a mega-mutation effectively 're-painted' its body and wings, changing it from orange and green to the striking black and yellow bee-like mimic that exists today.

The Instant Sparrow

A long time ago, there was a little brown bird that was hungry. It lived in the Middle East around the time the very first farmers started to cultivate their fields and put down roots. The birds watched the farmers tend to their crops and were quick to notice when grain and seed were accidentally split, then they swooped down and gobbled them up. The birds soon realised that it was easier to find food by hanging out with sloppy, wasteful humans, than it was by foraging in the wild. So when farmers began to spread out from the Fertile Crescent and take their technology elsewhere, the birds decided to follow them.

Around 8,000 years ago, these little brown birds – or Asian house sparrows as they are now known – followed the farmers into Europe, where they bumped into another little brown bird that was enticingly similar. The new bird, known as the Spanish sparrow, looked very similar to the Asian house sparrow, but it was a little bit bigger and a little bit heavier. Males sport a chestnut crown instead of an ash-coloured one and their cheeks are flashed with white rather than grey. The two species took a liking to one another and began to breed, and the offspring they produced sported a mix of characteristics, European on top and Asian down below; they had the facial markings of the Spanish sparrow and the body markings of the immigrant bird. Critically, the offspring were fertile when they grew up, and as time went by, the hybrids found they preferred to mate with each other rather than either of the parent species. A new species was born. It has since been dubbed the Italian sparrow, named after the Italian peninsula where the parent species first met.

Geneticists have since confirmed the species' mixed-race ancestry, and today Italian sparrows can be found in many parts of Italy and other regions of the Mediterranean. Evolutionary biologist Chris Thomas from the University of York has watched the little birds stealing crumbs from the tourists in the Piazza San Marco in Venice. Thinking about their evolutionary journey, Chris believes that the initial genetic changes would have been laid down during a single breeding season when the parent species first met. Sparrows are prolific breeders, raising multiple broods per year, so the new species could have become separate within just a few hundred generations or a few short decades. 'This is at least a thousand times faster than would be expected by our conventional view of evolution,' says Chris. 'It makes the emergence of the new species almost instantaneous.'

The story is remarkable. Here is an example of large-scale evolutionary change – the birth of a new species – that occurred rapidly because of hybridisation. Ten thousand years ago, the Italian sparrow was nowhere to be seen, and it only came into being because people invented agriculture and then exported the technology. 'The Italian sparrow exists because humans created the villages, farms and agriculture that enabled the Asian and Spanish sparrows to meet in Europe and create a new kind of hybrid,' says Chris. We humans are directly responsible for the genesis of the Italian sparrow.

Closer to Chris's home, there is another new hybrid species that has come into existence even more recently. Its story begins 300 years ago, when some well-meaning botanists brought a little yellow flower from the barren slopes of Mount Etna in Sicily to Oxford's Botanic Gardens in England. The plant, which was itself a naturally occurring hybrid, grew well in captivity but grew even better after its seeds escaped from the Botanic Gardens and blew around the city. They settled amid the cracks of the city's stone walls and the flagstones of the University's colleges, where the conditions were not so different from their volcanic homeland. One hundred years later, the plant had spread around the city, and was now so different from its two parent species that it was unable to breed with either of them. A new species had arrived: the Oxford ragwort, aka *Senecio squalidis.*

More was to come. On 12 June 1844, the city opened its railway station, providing passengers and plants with easy access to other cities. Seeds from the Oxford ragwort blew down the line, and put down roots in the gravelly track. As the plants grew and produced seeds of their own, the species began to commute away from Oxford towards other cities. It arrived in York in the 1870s where it met another member of the *Senecio* family, the native common groundsel. The two species hit it off, and a hybrid variety was born. A few decades later, the hybrid had become a self-perpetuating, genetically distinct bona fide new species known as Yorkwort, aka *Senecio eboracensis*, after the Roman name for York, Eboracum.

The situation then repeated itself again, in the mid-1890s, when the Oxford ragwort conquered North Wales. After colonising the railway tracks between Oxford and Wrexham, the scruffy invader interbred with the resident groundsel to produce a different hybrid, but this time speciation occurred even more rapidly. Remember that plants sometimes swallow whole sets of chromosomes in a process called polyploidy? Well, the Welsh groundsel, as it became known, appropriated a full set of chromosomes from both parents, giving it a grand total of 60. This was 20 more than the native groundsel and 40 more than the Oxford ragwort. As mutations go, it doesn't get any more 'mega' than this and were Goldschmidt still alive, he might even have called it a hopeful monster. The newly configured genome made the Welsh groundsel so genetically different from either of its parents, that speciation could have occurred the instant that the chromosomes combined.

Here then are three additional species that have all appeared in the last 300 years and that owe their existence to the actions of humans. 'Our activities are bringing together species that would never normally meet, and providing them with the opportunity to breed,' says Chris. Through our actions, we have become unwitting evolutionary matchmakers. Global warming is melting the Arctic, bringing polar bears into contact with grizzlies. Agriculture brought the Asian sparrow to Europe, where it spawned the birth of its Italian counterpart. And the British railway system blew the seeds of the Oxford ragwort into the arms of the native groundsel. Hybridisation is occurring as humans break down the geographical barriers that otherwise keep species apart. Most notably, as humans manoeuvre species around the globe, the opportunities for hybridisation increase.

Strangers in a Strange Land

In Scotland, the native red deer are hybridising with non-native Japanese sika deer. Introduced to the Highlands over a century ago, the sika escaped from the deer parks where they were initially kept and established themselves in the wild. They began breeding with red deer and produced young that were fertile. Although most matings still occur within species – red deer mating with red deer, and sika deer mating with sika deer – about one in 500 occur when red and sika deer interbreed. This is enough to mix things up. In some areas, up to 40 per cent of the deer are now hybrids, causing some to speculate that a new species is beginning to emerge.

Humans are transporting species around the globe at great speed. In the past, terrestrial creatures could only move between continents when the land masses physically collided; for example, when the world's disparate continents came together to form the supercontinent Pangaea 335 million years ago. Now, they travel on aeroplanes, boats, cars, bikes and even the soles of our shoes. Microbes, fungi, plants and animals are rapidly being transferred far beyond their natural range to the most distant corners of the Earth.

Invasive species, as we call them, have been the downfall of many an ecosystem, but for every horror story that exists there are far more happy endings. When non-native species cause trouble it tends to be because they are ecologically distinct from the resident flora and fauna. They are square pegs in round holes. The native wildlife of New Zealand, for example, is struggling to defend itself against invasive mammals, because it had no ground-dwelling mammals until 700 years ago. Elsewhere, however, where the differences are smaller, invasive species are less of a problem. When you look at the changes on a wider scale, the picture is less adversarial and more laissez faire. Invasive species don't *always* displace the locals. Sometimes they just settle down and fit right in. They get to work pollinating crops, fertilising the ground, spreading seeds and providing other useful ecological services.

It's interesting to consider, just for a moment, how irrational and inconsistent our attitudes to invasive species are. Sparrows, for example, spread around the globe from Asia. Now they are loved in the UK, where they visit our gardens, and inexplicably loathed in the US, where they are viewed as a nuisance. In the UK, there are more than 37 million rabbits and less than 1 million brown hares. Both were introduced by the Romans around 2,000 years ago, but while hares have become a national treasure, rabbits have become a national pest. We are less tolerant of recent interlopers, and more accepting of long-term immigrants. So the edible dormouse (a different species from the native dormouse), which escaped into the wilds of Britain from a private collection in 1901, is persona non grata in the UK, while the house mouse, which arrived during Neolithic times, is accepted as a native species. Meanwhile, the signal crayfish, imported to Britain from North America in the 1970s, is vilified for decimating our native white-clawed crayfish, even though there were already five other non-native species of crayfish present when it arrived, and even though our 'native' species has since been shown to be an immigrant.

Contrary to their bad-boy reputation, there are many instances where the arrival of non-native species has actually boosted biodiversity. A casual stroll into the countryside with my trusty GM wolf, Higgs, turns up a long list of foreigners. Depending on the time of year, rabbits and muntjac deer[*] dart for cover amid blooms of non-native plants such as bluebells, poppies, snowdrops, rhododendron and Himalayan balsam. In the autumn months, Higgs is taunted by grey squirrels[†] who wantonly cache their nuts just a hare's breath[‡] away from our path. When disturbed, they race into the boughs of non-native sycamore trees, leaving a frustrated and foolish mutt barking up into its canopy. We walk past cattle, sheep and horses[§] and fields full of wheat and barley[¶].

There are 1,875 established non-native species in Britain, yet as far as we know, they have never caused the global extinction of any native British species. In New Zealand, the 2,400 species of native vascular plant[**] have been joined by a further 2,000 non-native plant species. Far from stifling biodiversity, these foreigners have almost doubled the total number of plant species that now grow on New Zealand soil. Most co-exist peacefully alongside the native flora, with little sign they are damaging the environment. Despite what you might have heard, invasive species aren't *all* bad. Chris Thomas estimates that for every new invasive species that arrives, less than

---

[*] Introduced from China in the twentieth century.

[†] Introduced from North America in the nineteenth century.

[‡] Hares are non-native. So are their breaths. And yes, I prefer the phrase 'hare's breath' to its alternative, 'hair's breadth'.

[§] All introduced during the Bronze Age.

[¶] Domesticated in the Middle East's Fertile Crescent around 10,000 years ago. Now available in a supermarket near you.

[**] Vascular plants are the sorts of plants that most people recognise. They have a vascular system which helps them to move water and minerals around. Ferns, oak trees and sunflowers are all examples.

one native species disappears. Invasive species *can* do irreparable damage and there is sometimes a need to control them, but when the bigger picture is considered, invasive species actually help to raise biodiversity.

Hybridisation is part of this picture. 'We can be fairly confident that there has never been a time when so many species from geographically separate parts of the world have become mixed up so quickly,' says Chris. New connections are forming. Species are interbreeding. New varieties and species are being born. 'Speciation by hybridisation is likely to be a signature of the Anthropocene,' says Chris.

Our actions are fuelling evolution, and we should not be *too* sniffy about the creations that ensue. Conservationists bemoan the frisky Highland deer that have become 'contaminated' with the genes of an invasive species. Along with many other hybrids, the so-called 'Mongrel of the Glens' has been stigmatised for its lack of genetic purity, yet hybridisation is a natural part of evolution. *We* are hybrids after all. When hybridisation occurs in the wild, it's generally regarded as negative, yet hybridisation can provide a source of evolutionary novelty and a way of increasing genetic fitness. If nothing else, the sea-monkeys have taught us this. It's time we recognised that hybridisation is a valuable part of the story of life on Earth, and of humans' interactions with the natural world.

We need to tear ourselves away from the notion that life is a tree. In 1837, Darwin sketched a scruffy, bifurcating structure in one of his notebooks. A single black line split into three black lines, some of which split into more black lines, and so on. In the top left-hand corner of the page, he wrote the ponderous words, 'I think'. In the time that has since passed, that sketch has come to encapsulate evolutionary theory, but it suffers from a major flaw. At no point do twigs on the periphery of the tree loop back and meet with the limbs from other branches. If hybridisation is commonplace, which it is, there should be connections between branches. As well as genes being passed down discrete branches when members of the same species mate, they are also being transferred from limb to limb when disparate species hybridise. The diagram should be less of a tree and more of a bramble bush.

The sketch is neat, but life is messy. Living things don't care what taxonomic classification they have been given. They have no concept whether they are this species or that. Instead, the fundamental drive to reproduce guides their behaviour, sometimes with alarming ferocity. I once witnessed my friend's dog dry-hump her living-room curtains so vigorously that they nearly disintegrated. So should we really be so surprised when a polar bear falls for a grizzly?

CHAPTER EIGHT

Darwin's Moth

During the Second World War, the good people of London took shelter from one enemy, only to find themselves plagued by another. When German bombers blitzed the capital, Londoners fled to the bowels of the city and took refuge on the tracks and platforms of the Underground. Over the course of the war, some 180,000 people sought sanctuary in the dark, dimly lit tunnels that lie submerged beneath the capital's streets, but they were not alone. The standing water that collected underground from rainfall and cracked pipes attracted insects, and the evacuees soon found themselves scratching frantically and cursing this most insidious fifth columnist. Mosquitoes made their already difficult lives an absolute misery.

After the war, few people thought about the mosquitoes, until in the 1990s, a geneticist called Katherine Byrne decided to look for them. She joined maintenance crews as they travelled deep into the city's subterranean world, far beyond where regular commuters are allowed to go. The mosquitoes were still there. In blackened tunnels, she collected samples of their larvae from flooded sumps and shafts. Then she took them back to her lab at the University of London and analysed them.

The mosquitoes came from seven locations on the Central, Victoria and Bakerloo lines, and what she found surprised her. The mosquitoes from the three Tube lines were genetically different from one another. This was, she later explained, because the insects lived in three almost entirely separate worlds. As the trains came and went, they stirred up clouds of the insects but it was virtually impossible for different populations to mix. The only way a meeting could happen, she quipped, would be if they all changed trains at Oxford Circus. The three Tube line strains seemed to be evolving in separate directions, but they weren't just dissimilar from one another. Byrne also found that they were genetically distinct from the mosquitoes that lived above ground, in the city.

The local surface-dwelling mosquitoes had different DNA and lifestyles from their underground relatives. Above ground, the insects fed on birds rather than commuters. They hibernated in winter, mated in large swarms and required a blood meal before they could lay their eggs. More than 20 metres beneath them, the subterranean mozzies were active all year round, mated in confined spaces and were able to lay eggs before feeding. In recognition of the annoyance it caused during the Blitz, and to distinguish it from its regular surface-dwelling relative *Culex pipiens,* the underground population was dubbed *Culex pipiens molestus*.

With the benefit of hindsight it's easy to see how this new variety of mosquito came to be. After workmen finished building the Underground, in the mid-1800s, the tunnels were sealed off from the outside world and the mosquitoes that were down there became trapped.

There were no birds for them to feed on, so instead the insects began to feed on people and rats. They started mating in small spaces because they had no other option, and they stopped hibernating because the temperature was more or less constant all year round. Physically separated from each other for over 100 years, the two populations, above and below ground, had no opportunity to mate with one another, and when Katherine gave them the opportunity in a laboratory experiment, they were totally uninterested.

Over the course of maybe 100 generations, the Underground mosquitoes began to evolve along a novel trajectory. Their unique environments caused the populations to change, until they reached a point where they were so disparate, they could no longer interbreed. Now scientists are arguing over whether these differences are enough for *molestus* to be recognised as a separate species.

The London Underground mosquito is not a one-off. Humans are building cities, and cities are changing species. When we construct underground tunnels, lay roads and erect buildings, we radically alter the habitat and resources used by other living things. In 2007, the Earth reached a milestone when, for the first time in history, the number of people living in urban areas outnumbered those living in rural areas. By the middle of the twenty-first century, it's estimated that two-thirds of the world's projected 9.8 billion people will live in cities and towns. Wild spaces are being eroded, and wild things are exploiting new urban niches. Rats, pigeons, bed bugs and cockroaches have made cities their own. Lions have been spotted wandering through the slums of Nairobi. Wild turkeys are on the rise in Montreal and marmosets can be found living in the city parks of Brazil. To make themselves heard over the traffic's thrum, city-dwelling European great tits now sing faster, shorter, higher pitched songs than their forest-living relatives. Now a bird that used to sing like Pavarotti sounds more like Justin Timberlake. City life is causing animals to alter their behaviour, but a growing number of studies suggest that urbanisation is also having a much more profound effect. As the London Underground mosquitoes demonstrate, it is influencing the course of evolution.

In previous chapters, we've explored how the bustle of humanity is leading, on the one hand, to population declines and extinctions, and on the other, to the emergence of new hybrid creations like the Italian sparrow and the pizzly bear. But what of those species steering an alternative course through this melee of human commotion? All around us, species are evolving, but now human activity is forcing many to evolve at breakneck speed. This chapter is their story.

When Darwin laid out his theory of natural selection, he described the mechanics of evolution. Three key ingredients are needed. First, there must be variation. Members of the same species are all broadly similar, but crucially, key differences exist. Certain individuals, for example, may be larger or faster. They may differ in their colouring, in their ability to withstand drought or in their metabolic rate. Variation is the fuel that powers evolution. Second, there is selective pressure. Selective pressure is anything that affects an individual's ability to reproduce and pass on its genes. In Yellowstone Park, for example, wolves exert a selective pressure on Pronghorn antelopes because they hunt and eat them. Pronghorns enjoy eating tender green shoots, so the availability of these plants is another selective pressure on them. Third, sometimes a particular idiosyncrasy will give an individual the edge. Suppose there is an outbreak of disease that kills many of the pronghorn. The animals that are resistant to the disease are more likely to survive and go on to reproduce. Crucially, the genes underpinning this resistance will be passed on to the next generation. So the final ingredient in Darwin's recipe is heritability.

Environmental change is a source of selective pressure, and cities are a source of environmental change. Instead of earth and foliage, there is concrete and grass. There is light, noise and chemical pollution, not to mention traffic, roads and a general lack of green space. As their original habitats become squeezed, many species are turning to the cities, and in recent decades scientists have become fascinated with the changes that have ensued.

City Slickers

A few hundred years ago, the area that is now New York City was covered in forests and meadows. Lots of animals lived there, including a small, bright-eyed rodent called the white-footed mouse, which scampered through the grasses and the forest undergrowth. The mice moved freely, breeding with one another to create a single continuous population with a well-mixed gene pool. Then, from a tiny seed, the Big Apple grew. Skyscrapers shot up. Avenues and streets carved the emerging metropolis into neatly delineated blocks. Grassland gave way to houses, theatres and shopping malls. As its natural habitat was eroded, the mice sought refuge in the few green spaces that were left: the city parks.

Now isolated populations are thriving in New York's Central Park, Prospect Park in Brooklyn and smaller parks such as Willow Lake in Queens. Scientists who study them have found that, just like the mosquitoes from the various London Underground lines, the mice from the various parks have all evolved along slightly different paths. Now the mice all have their own park-specific DNA. Mice in Manhattan have a genetic signature that is distinct from mice from the Bronx, and different from mice from Queens and the Rockaway Peninsula.

When their genomes are interrogated more closely, the pattern of variation becomes even more revealing. Stephen Harris and Jason Munshi-South from the City University of New York compared DNA from urban and rural populations of white-footed mice, and found that the genomes of mice that lived in city parks show signs of becoming adapted to urban living. White-footed mice from Central Park, for example, carry unusual versions of genes that are involved in processing food and dealing with pollution. One gene, called *FADS1*, is known to help animals deal with fatty foods, while another, *AKR7*, helps to neutralise aflatoxin, a toxic compound produced by a fungus that grows on mouldy nuts and seeds. In the 120 years or so that the mice have been isolated in

Central Park, it seems as though they have been evolving the ability to eat New York staples such as pizza and peanuts!

Meanwhile, 2,500km (1,500 miles) away in south-western Nebraska, American cliff swallows face an alarming new challenge. In recent times, many of the birds stopped building their little round mud nests on crumbly overhangs and sandy cliffs, and started to build them on the undersides of highway bridges, overpasses and road-side culverts. These nest sites are structurally sound, but they end up perilously close to the fast-moving cars. Scientists Mary Bomberger Brown and Charles Brown have been studying these birds. Forty years ago, when they started their research, the wings of live swallows were the same length as the wings of swallows that died on the road. But by 2010, when they published their data, the wings of live birds were half a centimetre shorter than those of the unfortunate roadkill. In addition, there were far fewer swallow fatalities. The number of dead birds had declined by 90 per cent even though the traffic volume was unchanged. So what was going on?

The duo concluded that swallows with shorter wings were surviving because they were better at taking off vertically and dodging the oncoming traffic. Over time, these birds reproduced and passed the relevant 'short-wing genes' on to future generations, who found themselves at a similar advantage. Longer-winged birds, in contrast, were more likely to meet the blunt end of an SUV, so in time their 'long-wing genes' became purged from the gene pool. Natural selection, imposed by the dynamics of the urban setting, led to the evolution of smaller, more rounded wings and birds that can out-manoeuvre vehicles.

The more scientists look for evidence of urban-influenced evolution, the more they find. In Puerto Rico, city-dwelling crested anole lizards have evolved longer legs and stickier toes that help them cling to buildings. Orb-web spiders in Vienna are evolving to build their webs on the artificially lit handrails of footbridges, because moths are attracted to the bright lights. British great tits are evolving longer beaks because it helps them snaffle peanuts from bird feeders, while European blackbirds are becoming cockier because it helps them deal with the stress of city living. Plants are at it too. In Montpellier, a common weed that grows in pavement cracks – holy hawksbeard – is evolving heavier seeds that drop directly to the ground, where they can germinate, rather than lighter seeds that blow away and land on concrete. And so the list goes on.

Humans change the environment and life responds by evolving. When we build cities, we shape life in new and unpredictable ways. Urban spaces provide opportunities for individuals with the appropriate genetic wherewithal, but they're not for everyone. If a species cannot adapt to city life, or relocate elsewhere, then its days are numbered. When we build cities, we play with evolutionary fire, but urbanisation is only the most visible of humanity's selective pressures. There are many less obvious selective pressures that are having an equally profound effect.

Darwin's Moth

London's Natural History Museum is home to 11 million specimens of moths and butterflies from all around the world. Of this staggering number just a fraction are on public display. The remainder are stashed away, but if you ask nicely, Geoff Martin, the Senior Curator in charge of Lepidoptera, may let you take a peek.

I contacted Geoff because I was looking for a very particular moth; not just a particular species, but a particular individual belonging to that species. I last set eyes on this specific moth over 30 years ago, when I was a teenager. At the time, it was already pinned and mounted in a rather splendid mahogany display cabinet. The cabinet belonged to my great-uncle, Richard Pilcher, who lived in the wilds of Lincolnshire in a village called South Thoresby. When I knew him, he had retired from his career as a surgeon, and devoted much of his time to his second passion: natural history. Weekends spent with Great-uncle Rick were a treat. We'd bird watch, collect caterpillars and eat a lot of cake. When I could, I'd sneak into his garden room where the cabinet lived and examine its contents.

Over the course of his life, Rick had caught many of the UK's 900 or so species of larger moth. As was the custom with his generation of natural historians, he euthanised some of these specimens and built them into a collection. Shallow drawers with glass facades housed thousands of individuals, each meticulously prepared and stabbed through the thorax with a pin. There was drama and colour, subtlety and muted seasonal hues. Some were so big they could fill the palm of your hand; others were so dainty it seemed they would disintegrate if you so much as looked at them.

After he died in 1990, Rick's family bequeathed the collection to the Natural History Museum, and since then, Geoff and his colleagues have digitally catalogued each and every one of Rick's 4,577 moths. Now it's possible to turn up, ask to see a particular moth, and have Geoff show it to you.

We disappeared into the bowels of the museum and eventually found ourselves in a room full of rolling stacks, like a library. Instead of books however, the stacks contain thousands of trays of moths and butterflies.

'Let's dig out these peppered moths,' said Geoff amiably.

He disappeared purposefully and reappeared a few moments later holding a shallow tray full of monochrome moths. The moths were arranged, with fore and hindwings splayed, in a series of neat rows and columns. In this one tray alone, there were more than 150 individuals that morphed in tone from left to right like a colour palette for paint. On the left side, they sported sepia wings with dark dots and smudges. On the right side, they were almost entirely black.

'These are all peppered moths,' said Geoff. He explained how the moths were all from different collections. 'People leave them to us in their wills,' he said. 'Then we arrange them into trays containing particular species.' He inspected the tiny labels pinned next to the specimens. 'Richard Pilcher's is in the last column, fourth from the bottom.'

My eyes darted to the bottom right-hand corner, then I counted up. One, two, three … four. There was my moth: a sooty, skew-whiff specimen that I hadn't seen in years. Its wonky antennae fanned out at right angles to its furry black body, and its wings were dark and dirty-looking. It didn't look like much; not too big and not too small. There were no distinctive markings, or anything much to give away the fact that this moth is, in fact, an evolutionary show-stopper. The peppered moth has become the icon of evolution in action; the most famous example that there is of how city life and human activity can influence the trajectory of an entire species.

Before the mid-nineteenth century all the peppered moths in England were like those on the left-hand side of the tray: cream-coloured with black spots. Then, in 1848, a moth collector from Manchester caught and collected a sooty-coloured individual, just like Great-uncle Rick's moth. In the years that followed, this darker form became more and more common until, by the turn of the century, it had all but replaced the lighter form in Manchester and other industrial cities.

The cause was pollution. Coal burned in the early days of the Industrial Revolution produced soot that smothered the tree trunks where the peppered moths rested during the day. Blackened 'melanic' moths were perfectly camouflaged against this darkened background, while lighter moths stood out and were more likely to be eaten by local birds. Over time, because melanism is heritable, the number of sooty-coloured peppered moths increased in urban areas, and the number of lighter coloured moths declined. But the lighter moths didn't disappear altogether. Instead, they fared better in rural areas where the air was cleaner. Then when the Clean Air Act started to spruce up the skies in the mid-twentieth century, things changed again. The melanic form declined in urban areas, and the lighter form became more common.

Darwin would have been baffled. As I've mentioned before, Darwin thought that evolution happened slowly across immense geological time spans. He wrote in *Origin of Species*, 'We see nothing of these slow changes in progress, until the hand of time has marked the long lapse of ages.' He viewed evolution as a plodder rather than a sprinter, with changes happening so gradually they can never be seen in real time. He may have figured out how change occurs across generations, but he never seemed to imagine that this process could happen quickly.

The peppered moth was a conundrum. Here was a species that perfectly illustrated the process of natural selection – it even became known as 'Darwin's moth' – but its changes happened too fast. A moth that was one colour at the start of the Industrial Revolution was another colour by the end. In 2016, scientists deciphered the exact nature and date of the change that caused this dramatic transformation. A tiny piece of DNA, known as a transposon or jumping gene, had inserted itself into the middle of a gene that controls pigmentation. The mutation caused the moth, and all of its descendants, to develop anthracite-coloured wings. The date that the jumping gene landed: 1819. It took less than 100 years for the UK's peppered moth population to reinvent itself; less than a blink in the geological time frame that Darwin had envisioned.

When the story of the peppered moth emerged, its rapid transformation was considered an anomaly, but then biologists began to notice that other species were undergoing rapid changes too. Many of the changes were heritable, leading to speculation that evolution might not be the dawdler that Darwin had imagined.

In 1950, Australia was awash with imported rabbits, which were damaging crops and bothering the local wildlife. After much cajoling, the government decided to try a radical form of pest control: introducing more rabbits, this time infected with a lethal rabbit-specific virus called myxoma. The virus causes the infamous disease, myxomatosis, which kills rabbits within weeks. The bunnies were released into the Murray Valley of south-eastern Australia and from there, the virus spread. Two years later, 500 million bunnies had succumbed. And so began one of the greatest inadvertent experiments in natural selection that the world has ever seen.

Both the virus and the rabbits began to evolve. The original, highly lethal myxoma strain evolved into less deadly versions, buying the rabbits time to evolve resistance. Individuals with mutations that helped them survive the disease were able to multiply and pass these useful genes on to future generations. The rabbit population began to rebound, prompting the virus to evolve its own countermeasures in an evolutionary arms race that continues to this day.

Species evolve in response to environmental change. This is not unexpected. What surprised the scientists who studied the rabbits was the speed at which it happened. In one place, it took the rabbits less than a decade to evolve resistance to the myxoma virus. Scientists were beginning to realise that evolution can happen quickly. They just needed a way to test the idea experimentally. Guppies provided the answer.

In Trinidad, the little fish live in freshwater streams that trickle down the mountains via a system of waterfalls and pools. The waterfalls act as natural barriers, so fish that live above and below them can experience a variety of conditions. In some places there are lots of predators, while in others, there are relatively few. Back in the seventies, evolutionary biologist John Endler – now at Deakin University, Australia – noticed that in some locations, the guppies were vivid and spotty, while in others they were drab and colourless. In an attempt to understand why, he moved guppies from pools where there were lots of predators to pools that were almost predator-free. Two years later, when he returned, he found that the guppies swimming in predator-free pools were more colourful and spotty than those in predator-infested waters. They had changed enormously in a very short space of time, but it made sense. When predators are around, it's a good idea to blend in, but when they are absent it's OK to stand out, especially if the females of the species prefer gaudy bright patterns. Here was a clear example of rapid evolution in action. Endler was on his way to showing that male-guppy patterning is a trade-off between attracting females and avoiding predators, but there was more to come.

In the 1990s, American ecologist David Reznick performed similar experiments but this time he looked at the effects on reproduction. He moved guppies from a stream where they had to fend off aggressive cichlid fish to a second site where there were no predators, and he added cichlids to a pool above a waterfall where the guppies had previously lived in predator-free bliss. The guppies' worlds turned upside down. The guppies that now found themselves in danger started maturing earlier, while the guppies transplanted to peaceful waters started to mature later. This also makes sense. When predators are an everyday threat, life becomes about producing as many offspring as possible as quickly as possible, but when life is easy, this reproductive pressure is eased.

Reznick showed that invasive species – for that is what the cichlids were – had the potential to guide evolution along a different track. The study triggered a news story in the *National Enquirer* entitled, 'Uncle Sam wastes US$97,000 to learn how old guppies are when they die.' Reznick calls it one of his proudest career moments. Yes, mortality was monitored but the paper demonstrated something far more profound. The guppies' reproductive habits changed within just six to eight generations, or four short years. The rate of change was up to seven orders of magnitude greater than rates inferred from the fossil record. This wasn't just rapid evolution. This was evolution at warp speed.

Around the same time, other studies on different species in different settings were arriving at similar conclusions. Even Darwin's finches were at it. When studied closely, the beaks and body size of the very birds that inspired Darwin's theory of natural selection could be seen to shift across generations as the environment changed. The old paradigm began to dissolve. 'Darwin thought evolution was slow,' says evolutionary biologist Andrew Hendry from McGill University, Canada. 'Everyone believed him for 140 years, but now we know it's not the case. It can occur quite rapidly.'

In recognition of the fact that this rapid evolution occurs across human time-frames of years, decades and centuries, rather than the slow lapse of geological time, it has been dubbed 'contemporary evolution', and far from being the exception, it now seems the phenomenon is commonplace. As humans modify the world around them, species have no choice but to try to adapt as quickly as they can.

The Rise of Resistance

Today, pollution is an omnipresent threat. Vehicle exhausts, factories and power stations jettison microscopic particles into our atmosphere. Agricultural fertilisers, animal waste, detergents and sewage leak into our waterways. Between 1947 and 1976 the American multinational conglomerate General Electric dumped around 500,000kg (500 tons) of chemicals called polychlorinated biphenyls (PCBs) into the Hudson River. Back then, PCBs were widely used in industry as coolants, but the same characteristics that made them attractive to manufacturers also made them an environmental nightmare. PCBs take a long time to break down and are soluble in fat rather than water, so when they were dumped in the river, they accumulated in the fat stores of the crustaceans and fish that lived there, then passed up the food chain and were eaten by humans. This was not good. A growing number of studies have since linked PCBs to a wide range of health issues, including cancer and nervous-system disorders, and in the late seventies the chemicals were banned.

At the time, the effects of PCBs on local wildlife were not studied, and even though the river has since been cleaned up, many fish that live there now still carry high levels of PCBs. Atlantic tomcod pulled from the Hudson River have one of the highest PCB levels found anywhere in nature. By rights, they should be dead, yet remarkably, the Hudson River tomcod are doing swimmingly. They have evolved resistance to PCBs and now live and breed in waters that are still tainted. Further east, another species, the Atlantic killifish, has become immune to PCBs, dioxins and methyl mercury, and researchers are discovering species of fish that can deal with other pollutants too.

Our slapdash actions have led to their rapid evolution. In the space of just half a century – or a few dozen generations – tomcod and killifish have become resistant to toxins that would kill most fish. Unlike the peppered moth, however, these fish were not the lucky recipients of a well-timed mutation. Instead, they were able to draw from a healthy smattering of genetic variation that already existed. Scientists have found that the mutation that allowed the tomcod to survive was around at least 10 years before General Electric began polluting the river. Certain fish carried the mutation, but until the water quality began to deteriorate it offered no selective advantage. Then, when PCBs came along, these fish found they had an edge. They were more likely to pass on their genes than fish that lacked the mutation. As a result, resistance spread.

This standing genetic variation, as it is called, is thought to underpin many instances of rapid evolution. Increasingly, it's being seen in species that are small, numerous and fast breeding. Bacteria are becoming resistant to antibiotics, and agricultural pests are becoming immune to pesticides.

Aphids are a real problem. The small, sap-sucking insects eat and weaken agricultural crops, and transmit harmful viruses. They're also prolific reproducers. Although they can reproduce sexually, often they forgo sex and make clones of themselves instead. In a single year, one aphid can give rise to 18 generations without ever needing to mate. If all of those aphids were to survive and reproduce, at the end of the year we would find ourselves smothered in a blanket of aphids 150km (93 miles) deep.

If you like your blankets made of wool rather than aphids (who doesn't?), you'll be glad to know that this nightmare scenario is unlikely to happen. If they don't get eaten, aphids live for a few weeks, then die and deteriorate. They never build up to blanket proportions, but their phenomenal powers of reproduction mean two things. First, aphids are more likely to generate useful mutations than insects that reproduce more slowly, and second, these useful mutations can then spread through a population like wildfire.

At Rothamsted Research in Harpenden, England, entomologist Steve Foster hoovers up aphids in giant upside-down vacuum cleaners. 'They suck in aphids, beetles; anything that's flying really,' says Steve. 'Occasionally we've caught bats and birds. One time we even caught a ham sandwich. I think a bird might have dropped it in, or a student might have thrown it up there.'

What Steve and his colleagues have found is that aphids are becoming increasingly resistant to insecticides. 'One year, you can have very low levels of resistance,' says Steve, 'then the next year, 80 to 90 per cent of the aphids that you see all have that form of resistance.' First they evolved resistance to organophosphates and then, after those chemicals were banned in the UK, the aphids became resistant to their replacement, the pyrethroids. 'Now almost all of the peach potato aphids we collect are resistant to pyrethroids,' says Steve. 'The problem is that the insects just keep fighting back.'

When fish evolve to swim in polluted waters, we are happy for them, but when insects evolve resistance to pesticides, we are less than impressed. They are a threat to food security and to farmers' livelihoods. Pesticides are a particularly intense form of selective pressure. When we douse our crops with insecticides and poison our pests with chemicals, we promote the evolution of resistance, but every time we hit them, they come back stronger. Now scientists are concerned we are running out of options. When designing new insecticides, researchers develop chemicals that target key proteins, but there are a finite number of these targets. 'There are only a few ways to kill insects,' says Steve. 'We've gone through a lot of those ways. I think we're running out of compounds that work.'

We are locked in an arms race between pests and pesticides. Almost as fast as chemical companies develop new insecticides, aphids develop resistance to them. It's a similar story elsewhere. Human body lice, bedbugs, sand flies and black flies are becoming increasingly difficult to kill. Mosquitoes are evolving resistance to DDT and rats are becoming immune to over-the-counter poisons. More than 200 weeds have evolved resistance to at least 150 herbicides, and more than 500 species of pest have become resistant to one or more type of pesticide. Every time we come up with a new pesticide or herbicide, the pests and plants they are designed to target evolve in response. *We* are driving their rapid evolution, but it's not only small, fast-breeding species that react in this way …

Bighorn Mountain

Large, slow-breeding animals are also evolving rapidly as a result of our actions. On the aptly named Ram Mountain in Alberta, Canada, bighorn sheep are sought after by trophy hunters. The sturdy beasts sport spectacular curved horns which some people, inexplicably, use to decorate their homes. Bigger, apparently, is better, so hunters tend to target the rams with the biggest headgear. As a result, over the last 40 years, horn size has shrunk by more than a fifth.

Of course, if you shoot all the big-horned individuals there will be relatively more small-horned individuals left, but there is also something more fundamental going on. In Alberta, hunters need a licence to shoot the rams and are only allowed to kill individuals whose horns have reached a particular length: the so-called 'four-fifths curl' where the tip of the horn is level with the eye. The problem, however, is that at this point in their life, the animals have yet to reach the peak of their sexual prowess. So individuals that would have enjoyed high mating success at around eight to ten years old, are being killed at four to seven years old. They don't get to mate and they don't get to pass their 'big horn genes' on to the next generation. As a result, trophy rams are becoming rare and the number of licences issued now vastly outnumbers the legally available rams.

Similarly, elephants are evolving smaller tusks and sometimes, no tusks at all. In Zambia's South Luangwa National Park, the proportion of tuskless female elephants has increased from 10 per cent in 1969 to almost 40 per cent in 1989 as a result of ivory poaching. There's a rather satisfying irony here. Hunters always want 'the biggest': the biggest horns, the biggest tusks and the biggest animals. But it's a short-sighted strategy. The selection pressure caused by hunting causes a long-term reduction in size. Our lust for the large and the impressive is leading to the emergence of the small and the mediocre. Long live the underdog!

Things are no less dramatic in our oceans, where heavily exploited species are also reproducing earlier and becoming smaller. Salmon, cod, grayling and sole are just some of the species that are affected. Pacific salmon have shrunk by a staggering 25 per cent across just 20 years, while Atlantic cod that used to mature at around eight years of age are now becoming sexually mature in half that time. It's a real-world version of David Reznick's guppy experiment, but instead of cichlids predating the fish, it's us.

Humans are a predator like no other. When we hunt, we act in a uniquely human way. Where normal predators target the weak, young and puny, we target the strong, old and impressive. Normal predators aren't wasteful. Most of the time, they kill only what they need to survive. In contrast, we are bloodthirsty and excessive. We have the ability to kill in vast numbers, influencing not just the fate of the animals we destroy, but the evolutionary fate of entire species. It's estimated that hunting is causing the characteristics of many species to evolve at three times their natural rate, and sometimes we don't even need to catch or kill the animals that we hunt in order to have an effect.

The largemouth bass is a big olive-coloured fish found in the freshwater lakes of North America, where it is highly prized by anglers. The feisty fish are famous for putting up a fight when they find themselves on the end of a lure. They thrash and flail, making landing them a sweaty, testosterone-fuelled affair.

Studies have shown that the characteristic fishermen value the most – the willingness of a bass to take a lure, showing its aggressiveness – is a highly heritable trait. It makes sense then, that over time, the number of aggressive bass has declined and the number of calmer bass has increased. Only that shouldn't be the case, because bass fishing is predominantly a 'catch and release' sport. The aggressive fish *are* being caught, but they're not being killed. So why is it that, in some lakes, aggressive bass are in decline?

Part of the reason is that fish that are repeatedly caught and released are more likely to die early than those that evade capture. Stress takes its toll and as a result, the aggressive fish have fewer offspring. Parenting also plays a role. Male largemouth bass are diligent dads that guard their newly hatched offspring for several weeks. Aggressive dads are better at fending off predators, but they can't guard their fry while they are on the end of a fisherman's line. This gives predators a chance to nip in and gobble up their youngsters, and as a result, aggressive dads have fewer surviving offspring. Calmer fish, in comparison, are less likely to get hooked and so are better at defending their young, ultimately making them more successful at passing their genes on to future generations.

The practice of catch and release is well motivated. The idea is that it gives breeding females more of a chance to mate and produce offspring, but it doesn't consider the repercussions for the males who are negatively affected by the process. It seems that even when we deliberately try not to have an adverse effect on the environment, sometimes we have an adverse effect on the environment.

The same may be true in the world's fish hatcheries. Hatcheries breed and then release billions of wild juvenile fish every year. The practice has good intentions. Depending on the species and location, sometimes these fish are used to top up naturally declining populations and sometimes they are caught for food, but the underlying assumption is always that captive breeding gives the juveniles a decent head start. Unfortunately, this is not always the case.

Scientists have studied what happens after wild steelhead trout are brought into captivity, allowed to breed, and then their offspring are set free. When the offspring return to spawn, they produce 40 per cent fewer offspring than their captive-raised parents. 'It's a massive decrease,' says Michael Kinnison from the University of Maine, who studies contemporary evolution. Even when the fish are kept in conditions designed to be as natural as possible, their descendants still struggle on release.

'It makes sense when you think about how different the human environment is to the wild environment,' says Michael. Selective pressure, remember, is anything that affects an individual's ability to pass its genes on to future generations. Subtle, almost imperceptible differences can act as sources of selective pressure. Maybe the water is a slightly different temperature or the nutrients in the fisheries are a little bit different. It doesn't matter that the changes are small. If they are persistent, the repercussions can be large. 'Even if you're not actively selecting to improve these organisms in captivity, there can still be very strong selection when they go out the door,' says Michael.

It's actually extremely difficult to generate an environment that is *not* selecting. It's easy to imagine selective pressure as something we humans can turn on and off like a tap, but this is misleading. 'Selection is a process and it's a balance. It's a balance between all of the challenges that organisms face in their lives,' says Michael. 'The consequence is that if you try to make an environment that is not selective, you simply end up shifting the balance. You don't really remove selection altogether. You weaken the selection that's happening in one part of the system, freeing things up so that evolution by selection can happen elsewhere.'

It's quite a responsibility. If we catch fish, we affect their evolution. If we throw them back, we affect their evolution. The same is true if we feed birds or stop feeding birds, pollute rivers or clean them up, increase or decrease atmospheric carbon dioxide levels, and warm or lower global temperatures. The magnitude of human influence is now such that whatever we do, we change *something* and nudge evolution along an altered course.

As our numbers have grown, our impact on the planet has intensified. Now humans have become an evolutionary force of extraordinary influence. Human activity is a ubiquitous source of selective pressure. Our actions are pushing the evolution of living species into overdrive. New characteristics are emerging and life is changing right in front of our eyes. We steer evolution when we make grand, collective actions like building cities, burning fossil fuels and plundering the oceans, but we also effect change through smaller deeds, like feeding the birds or fishing for bass. We have unleashed a global tide of rampant evolutionary change, but what will the consequences be?

On the face of it, contemporary evolution seems to offer a lifeline to species that are being battered by the Anthropocene. Evolution is the mechanism that enables living things to adapt to a changing environment, but it doesn't plan ahead. There is no forethought or grand design, so an adaptation that helps an organism cope with the here and now is no guarantee of future success. In addition, contemporary evolution often comes at a price. The problem is that all organisms have a finite supply of energy, so when resources are diverted to help fuel some new feature, it leaves less energy available for other important processes. Aphids that have evolved resistance to certain pesticides, for example, are sometimes less good at dealing with the cold. They might survive the spray gun only to be killed by the frost. Antibiotic-resistant bacteria are sometimes less mobile. They might survive the prescription medication only to find they are less able to spread. Sometimes contemporary evolution is a game of swings and roundabouts.

As we teeter on the brink of mass extinction, it's not always easy to predict which species will succeed and which will go the way of the dodo. If the rate of environmental change outstrips the ability of a species to adapt and evolve, then its future is gloomy. Life's losers are likely to include the large, slow-breeding animals, as well as populations of organisms that are small and/or have limited genetic diversity.

The winners will be the ones that can reinvent themselves and adapt to life in our rapidly changing world. As these adaptations and their genetic underpinnings stack up, new species will emerge. The London Underground mosquitoes, the mice in Central Park, the pesticide-resistant aphids and the pollution-resistant tomcod – in time, they could all morph into distinct new species. Contemporary evolution is fuelling a wave of speciation. New twigs are starting to sprout on the evolutionary bramble bush of life. It may not be obvious to us now as we watch in real time, but it will be evident to future geologists who look back and study the fossil record from this period of time.

It's hard to quantify the rate at which humans are driving speciation. When species form by hybridisation it's easier to pinpoint their moment of conception, but when species arise via the gradual accumulation of genetic change, it's impossible to define the exact moment when speciation occurs. Here, speciation is an ongoing process rather than an event. Classically, biologists used to imagine that new species take hundreds of thousands or even millions of years to arise, but in the Anthropocene, new species are emerging across thousands of years or less. This change is not going unnoticed.

When evolutionary change occurs, it never happens in a vacuum. When one species adapts and starts to evolve, there are always repercussions. All living things are integral components of the ecosystems that they live in. 'Sometimes you might see a little bit of contemporary evolution that might not seem like much,' says Michael, 'but then you see how it fits into a food web ...' Tweak one part of an ecosystem and it can trigger profound changes in other parts of the ecosystem.

One hundred years ago, for example, cod were apex predators. The impressively streamlined fish hunted haddock, sand eels, crabs and other ocean critters. Seventy years ago, humans began to hunt them on an industrial scale. Super-trawlers caught hundreds of thousands of tonnes of cod, until 30 years ago, the industry collapsed. The cod population was battered and those that managed to evade capture experienced rapid evolution. In a race to reproduce before they were caught, the fish started maturing at a younger age. They evolved to become smaller. All of a sudden, a species that used to swagger at the top of a food chain found itself relegated to the bottom. Predator became prey, and the ecosystem changed.

Off the coast of Nova Scotia in Canada, nearly all of the large cod disappeared. As a result, the smaller fish and crustaceans that they preyed on became more common. This had knock-on effects too. The large-bodied zooplankton that the small fish fed on became less common through predation, and numbers of the teeny tiny phytoplankton that the zooplankton ate increased as a consequence. Phytoplankton are the base of many aquatic food webs. They use photosynthesis to convert sunlight to energy, and along the way mop up organic compounds like nitrates and phosphates. As phytoplankton became more common, nitrate levels decreased. Over-hunting altered the evolution and ecology of one of the ocean's top predators and ultimately changed the chemical composition of the ocean.

In this case, a temporary ban on cod fishing had a positive effect. Moratoriums introduced in the early nineties gave the cod a much-needed break. Although the recovering fish are still lighter than they were historically, stocks have been increasing. We have managed to turn things around, but we can't always presume that it's easy or even possible to undo our mistakes.

It's difficult to predict what the long-term repercussions of our actions will be. Lizards are evolving stickier toes. Moths are changing colour, fish are shrinking and bighorn rams don't quite live up to their name. At first glance, these changes might not seem important. When I worked at the science journal *Nature*, we had a phrase for those bits of scientific research that were novel and quirky but without obvious, serious implication. We referred to them as 'ooh, there's a thing'. It would be easy to dismiss these little bits of contemporary evolution as 'ooh, there's a thing', but that would seriously undermine their significance. Evolution, ecology and environmental change are irrevocably intertwined. In recent decades we have become more aware of the rampant evolutionary change that is occurring on our watch. What we need to be more aware of is the ecological and environmental fallout that will almost certainly follow.

CHAPTER NINE

Resilient Reefs

Sometimes great things happen in the most unlikely of places, such as a tiny, cramped corridor in a south London basement. The world's coral reefs are in trouble. We all know that. In recent decades, up to half of the planet's coral reefs have been all but annihilated. Their only crime? Evolving too slowly to keep up with the pace of climate change. But I've heard of a man who works in a hallway, whose research is helping to save them. He's taking evolution into his own hands – literally – and helping coral to reproduce using in vitro fertilisation (IVF). My mind is bursting with questions. *Where* does he get the 'raw material' from? *How* does one perform coral IVF? And *how* do corals, those famously static sea-dwellers, have sex in the first place?

In the next few chapters, I'm turning to conservation and looking at some of the extreme and ingenious methods being used to safeguard non-human species, and fuel life. When biologists 'conserve' living things, they are deliberately shaping evolution. Sometimes they are addressing a decline in abundance. Sometimes it's a loss of genetic diversity. Often it's both, but by managing wildlife and by dreaming up clever ways to redress these key issues, conservationists are creating brighter futures for endangered species and the world at large.

Thousands of years ago when we domesticated animals, and later on as we selectively bred them, we did it for our own benefit. The technology was for our gain, to produce useful resources like food and beasts of burden. When the molecular tools to genetically modify organisms came on board, we tweaked these species further, but it was almost always for our own ends. We have, for example, fashioned fast-growing salmon and meatier pigs. We've made animal models of disease, glow-in-the-dark fish and edited mosquitoes to curb the spread of malaria. It's a self-centred viewpoint that prioritises the needs of our own species over that of others, and it's desperately myopic. There have been five mass extinctions during the Earth's history. These have been caused by natural events like volcanic eruptions and asteroid strikes. Each time, 50 to 95 per cent of the world's species disappeared and each time the world took many tens of thousands of years to recover. Now scientists believe we are sleep-walking into a sixth mass extinction, caused by the actions of humans.

Conservationists seek to stem this tide of loss by protecting wild species and their habitats. The focus is being turned away from humans towards the wild species that conservationists now care for, with the recognition that if we can bolster biodiversity, it will be good for the environment and good for us too. This is a win-win strategy.

This chapter is about coral: an elegant, vibrant and ecologically crucial group of animals that live in dangerously warming waters. They are the subject of intense conservation efforts that focus on helping the coral to evolve. Jamie Craggs, who manages the Horniman

Museum's public aquarium, has spent much of the last decade finding ways to persuade captive corals to reproduce using IVF. I've heard of researchers who are deliberately speeding up evolution and building reef-saving 'Super Corals'. I want to find out more, so I email Jamie and ask if I can visit. I'm intrigued to see both his coral and where he works.

The response is enthusiastic. My timing is unexpectedly prescient. Next week, something big is happening. Jamie's corals are expected to spawn: the process whereby they release vast quantities of eggs and sperm into the water. If I'm there on Wednesday, between 1.00 and 1.30 p.m., I should be able to witness the spectacle. After that, he'll be performing IVF, so I could get to see the beginnings of a new reef being made.

That Jamie can predict the spawning time so precisely seems little short of coral clairvoyance, but he remains confident. 'If you come then, you'll get to see everything,' he says. So I book my train ticket and wait.

There have been corals in the world's oceans for over 400 million years. They belong to the group Cnidaria, a varied assortment of some 10,000 aquatic species that include sea anemones and jellyfish, which all have specialised stinging cells. Unlike sea anemones and jellyfish, however, corals live in colonies. Each individual coral is made up of thousands of much smaller, genetically identical organisms called polyps. Each polyp is essentially a tube-like stomach topped by a tentacle-fringed mouth. The tentacles sport specialised cells called nematocysts, which contain a venomous coiled thread that can be launched like a missile and used either to catch food, such as plankton, or to attack neighbouring corals that are encroaching on their patch. They may be static, but corals are feisty.

Sometime after corals evolved their basic body plan, they acquired a new addition that persists to this day. Corals contain algae called zooxanthellae that live in their cells and give the coral its colour. Algae are plant-like organisms that use photosynthesis to convert sunlight into organic carbon products, such as glucose and amino acids. During the day, this provides the coral with most of its energy needs, and in return, the algae get a nice place to live, with a stunning sea view. Algae living inside an animal: it's a beautiful example of symbiosis where two organisms live together for mutual benefit.

There are more than a thousand different coral species, conceived in every colour of the rainbow, from Barbie pinks to citrus limes and gaudy neon yellows. Some corals have large branching fronds; others have flattened, plate-like surfaces. Some are compact, with convoluted in-foldings, like a human brain, while others are long, thin and lanky. Some have been compared to lettuce leaves, others to candy canes. There are soft-bodied corals and hard-bodied corals, but only the latter build reefs. The polyps of reef-building corals secrete a rigid skeleton of calcium carbonate, which cements them to their neighbours and builds up over time to create the reef's structure. We may like to compliment ourselves on our architectural skills, but these are nothing compared to the corals' inventions. Reefs are infinitely complex. They are the largest animal-made structures on the planet. The daddy of them all, Australia's Great Barrier Reef, is so big it can be seen from space. Stretching 2,300 kilometres (1,430 miles) along the Queensland coastline, it covers an area the size of 70 million football fields, all made by individual polyps no bigger than a pea.

The Horniman Shuffle

When the big day arrives and I board the train for London, I leave behind a wintry landscape. Two days ago, central England was blanketed in snow. At night the temperature dropped to well below zero, so I wrap up warm with lots of extra layers. But as the train heads south, I feel the temperature rise and watch the snow melt away. I'm happily cosy and excited for what the day may bring.

In London, I trudge up the hill from the train station and turn into the museum grounds. It's pleasantly familiar. The Horniman Museum, with its eclectic collection of natural-history curios, was a favourite haunt when I used to live nearby. I'm met at the front desk by Jamie, a tall, energetic man with a broad smile and a hipster beard. He tells me that everything is going to plan and that the coral will shortly perform.

We trot down a flight of stairs into the bowels of the Museum, through a locked door and into Jamie's world. Ably assisted by a team of three researchers, this is where he breeds and cares for his coral.

'This is where our spawning tanks are. We're a bit pushed for space,' he says apologetically.

He's not kidding. We walk in single file through a narrow corridor that is lined on either side by large, long aquarium tanks. The air is full of the sound of whirring motors, humming filtration units and gurgling water. Some of the tanks – the ones used for spawning – are covered in blackout blinds. In the wild, coral spawning is exquisitely sensitive to light, as well as many other factors. Jamie tells me that in Singapore, for example, spawning happens in April, five days after the full moon, and begins at around 9.10 p.m. In Florida, it occurs in August, three to five days after a full moon, and starts around 10.45 p.m.

Microprocessors control the conditions inside the tanks so with the press of a button, the coral's environment can be tweaked to mimic that of its wild relatives. It's because of this, and five years' graft figuring out the exact parameters that trigger spawning, that Jamie can confidently predict the time that his corals will release their eggs and sperm.

Along with the vast majority of reef-building corals, the species here are broadcast coral, so-called because the various colonies spawn in synchrony, broadcasting billions of spawn into the water at the same time. Unlike most other animals, corals can't get up and physically move around. This makes it difficult to find a mate and reproduce. Synchronous broadcast spawning is the answer. The currents slosh the spawn around, so although coral is immobile, the spawn from one individual can meet and mix with the progeny of another. Jamie's corals are also hermaphrodites, meaning that each and every polyp produces both eggs and sperm, which are released together in discrete bundles.

I'm allowed to peek behind the blackout blind. Under the eerie red light of a head torch, I gaze into the tank and see hand-sized patches of coral growing on a grid-like floor. These are *Acropora millepora*, a branching species of coral that Jamie has been allowed to import from the Great Barrier Reef. In the wild, they come in shades of pink, green and brown, but today, the darkness and artificial light make it hard to work out their coloration.

An hour or so before spawning begins tiny pink spots appear all over the surface of the coral. The spots are bundles of eggs and sperm, forced up into the mouths of the individual polyps. It gives the colony a knobbly appearance. It's called 'setting' and is a sure sign that spawning will soon begin when the bundles are released into the water.

Things are looking good, so we walk down the corridor to the lab where the IVF will occur. We have to turn sideways to squeeze past one of the other researchers.

'It's called the Horniman shuffle,' says Jamie with a smile.

The lab is equally bijou, about the size of a downstairs toilet. On the worktop there is a sink and a microscope, and a suspiciously cloudy half-pint of liquid.

'That's a beaker of coral sperm from yesterday,' says Jamie, matter-of-factly. It's a combination of words that I never expected to hear organised into a sentence. Ever! Yet Jamie tosses it into the conversation as if a leftover glass of coral sperm is the most natural thing in the world.

I'm too polite to say, but am beginning to notice that the small room we're in is incredibly hot. I'm starting to regret the thermals that I put on when I left my freezing home. As if reading my mind, Jamie offers an apology.

'We have to whack the radiator right up so the sperm and the eggs don't get cold while we're doing the IVF.' There's also a large sheet of black plastic hanging in the doorway. 'To prevent heat loss,' says Jamie.

It's extra busy this week as the team have welcomed a new member, Keri O'Neil, Coral Nursery Manager from the Florida Aquarium's Center for Conservation (CFC). Keri is visiting so she too can learn how to become a coral clairvoyant. Researchers have studied captive coral spawning for a long time, but they have never been able to alter the event's timing. With his clever computer programs, Jamie can persuade his coral to spawn whenever he wants. At the Horniman Museum, spawning has become so predictable that the team can plan their tea breaks around it. Keri wants to learn how to control coral spawning in the lab, so she can take the technology home and use it to help save the native coral in Florida.

Blitzed by Bleaching

Corals live in tropical waters around the globe. Although their reefs occupy less than a tenth of 1 per cent of the ocean floor, they house 25 per cent of all known marine species. Millions of species call them home, making coral reefs one of the most biodiverse ecosystems in the world. Reefs are like underwater cities, providing homes for everything from sea cucumbers and tropical fish to cuttlefish and crustaceans. If we lose the reefs, we are in danger of losing the life forms they support, creating ripples that will spread far beyond the water's edge.

On land, hundreds of millions of people depend on coral reefs for their food and income. The reefs are a draw for tourists and also act as natural breakwaters, protecting thousands of miles of shoreline from coastal damage. In total, they provide goods and services worth at least £27 billion (US$34 billion) per year, but if they disappear, the annual expected damages from flooding would balloon to an estimated £217 billion (US$272 billion).

Keri has witnessed the demise of her local coral reef first-hand. The Florida Reef Tract stretches for 270 kilometres (170 miles) along the eastern coast of the US. It is the third largest in the world, after the Great Barrier Reef and the Belize Barrier Reef. Recently, scientists compared the latest satellite images of the region with historical charts and found that, in the last 250 years, half of the reef had disappeared. Gone! Just like that! Where there used to be thriving coral colonies, now there is only mud and scattered sea grasses. Things have become considerably worse in the last few decades. 'There are areas of the Florida Reef Tract that I've seen in my lifetime go from 20 to 30 per cent coral cover, to only 1 or 2 per cent. There are some areas now that I don't even recognise from 10 years ago,' she tells me.

Natural forces, such as tropical storms and heatwaves, are partly to blame, but as with so many woeful wildlife stories, the role played by humans is damning. In the last few hundred years, the population of Florida has rocketed from tens of thousands to well over 20 million, making it the eighth most densely populated state in the US. Cities have been built, giant causeways created. Pollution has increased, and the chemicals used to treat crops and lawns inevitably run off the land into the ocean. Meanwhile, the waters are becoming warmer and more acidic. Corals and their resident algae are exquisitely sensitive to these changes, and when the going gets tough, the algae get going.

Bleaching occurs when the algae become stressed and photosynthesis is compromised. This leads to the production of toxic molecules called reactive oxygen species, which leak into the coral tissues, triggering a cascade of molecular events that culminates in the algae being ejected from the coral. The algae are expelled into the water, turning the coral a ghostly, translucent white colour. Although coral can recover from short bouts of bleaching and set up new symbiotic relationships, major bleaching incidents are becoming more common. In the early 1980s, they occurred about once every 25 years, but now they happen once every six years. Widespread, mass bleaching events, which were unheard of 40 years ago, occurred in 1998, 2010 and most recently, between 2014 and 2017. This last event was the longest, most deadly coral-bleaching event in recorded history. Unusually warm water spread around the planet, devastating reefs everywhere. Over a two-year period, it killed off half the coral on Australia's Great Barrier Reef.

That such a blow should befall a World Heritage Site, protected under legislation for over half a century, is devastating. Corals *can* recover from short bouts of bleaching, but they need time to do so and time is running out. They regrow too slowly to keep up with the current pace of destruction. Bits that break off can re-attach to the rocks and kick-start a new colony, but that doesn't help to mix up the gene pool. To do that, corals need to be able to reproduce sexually. Genetically distinct colonies need to be able to mix and match their sperm and eggs, but in some places this is challenging because the surviving colonies are now so isolated that there's little chance their sperm and eggs will ever meet the spawn of a non-related colony.

'I feel like in Florida, to all intents and purposes, the fertilisation rate is zero because the colonies are just so far apart,' says Keri, as Jamie ducks out of the lab to see if the spawning has begun. 'That's why this intervention is so important.'

He returns a minute or two later with a grin a mile wide. 'It's starting,' he says. 'Come and see.' I follow him up the corridor and pry behind one of the large, black plastic sheets. Inside the tank, the 'spots' have erupted and hundreds of tiny pink balls, each no larger than a granule of sugar, are being released into the water. Each ball contains maybe eight eggs and tens of thousands of sperm all packaged up together. The eggs contain buoyant lipid molecules, which cause the bundles to float to the surface. I look at my watch. Jamie the Reef Oracle was right. It *is* between 1.00 and 1.30. It is 1.15 p.m. exactly.

The cramped corridor and tiny lab erupt into a frenzy of activity, as Jamie and the rest of the team tend to the perfectly timed arrivals. Over the next half-hour, colonies from three different *Acropora* species release their spawn, which are scooped out of the water in plastic half-pint glasses and carried carefully into the toasty lab. The contents are then stirred, a move that mimics the action of waves and natural underwater currents. It prompts the bundles to break apart into their constituent components, creating a murky half-pint of spermy water topped by a 'head' of minuscule pink eggs.

After that, the two layers are physically separated when Jamie pops a long 'straw' into the mixture and sucks from the other end to draw the liquid through. The milky sperm mixture flows into a second beaker, leaving the floating egg layer behind. In vitro fertilisation occurs after the eggs from one colony are poured in with the sperm from another. Egg meets sperm and the two combine. A few hours later and the fertilised eggs have started to divide. New life is beginning. A single beaker can contain many thousands of successful fertilisation events, but Jamie is modest. 'It's just bucket chemistry,' he says.

After 24 hours, the embryo reaches what Jamie calls the 'cornflake stage', a technical term where the developing coral resembles a tiny flake of puffed breakfast cereal. After that, it approaches what Keri calls the 'prawn chip stage', another technical term where the slightly older coral does an impression of a savoury snack. A few days later and the embryo has become a 'planula', a disappointingly named stage coined from the Latin word 'planus', meaning 'flat'.* The embryonic coral is now a slightly squidged, free-swimming larva that is transferred to a new tank where it swims its way to the bottom and attaches itself to the floor. A short while later, Jamie will add in the symbiotic algae that the developing corals need to grow. After that he'll feed them, look after them and watch his corals grow up.

The Bearded Pole-vaulter

In Victorian times, people marvelled at coral reefs and the myriad life forms that lived in and around them. In keeping with the socio-political culture of the time, they viewed the underwater ecosystem allegorically. Just as

* I'm lobbying for this to be called the 'naan bread stage', but the idea has yet to gain traction.

the poor workers toiled in crowded factories, so too, the cramped polyps laboured, building their reef for the benefit of a greater good. Society as a whole benefited from their actions, making corals the epitome of harmonious living.

At the same time, however, they realised that corals had a more sinister aspect to their nature. The steep sides and jagged contours of coral reefs made them like giant, underwater icebergs. In a time before GPS and ocean-floor mapping, ships often didn't see reefs until it was too late and the hull had been pierced by a sharp shard of coral. Ships frequently ran aground on coral reefs, so when Darwin set sail on HMS *Beagle* in 1831, one of his goals was to map the distribution of coral reefs. An obsessive hoarder of natural specimens, he broke off bits of coral to add to his stash, but collecting them was no picnic. Scuba diving had yet to be invented and Darwin didn't like to get his toes wet, so instead of approaching the corals from the water, he decided to approach them from above. At the Cocos (Keeling) Islands in the Indian Ocean, the top of the reefs are exposed at low tide, so Darwin waited for this moment, then picked his way across. Just as the British shoreline is pitted with tidal pools and crevices, so too the surface of a coral reef is pockmarked with hidden gullies and sharp edges. The surface was treacherous, so Darwin used a 'leaping pole' to propel himself from one piece of firm ground to the next. I like this image. It paints a very fallible, human picture: the father of evolutionary theory with his trousers rolled up, pole-vaulting across a coral reef. I wonder what the great man would have made of the state of our oceans today, or, for that matter, the work of Jamie in his south London basement.

In the past when scientists wanted to do coral IVF, they had to swim into the ocean and scoop up the sperm and eggs that they needed in special nets. 'It sounds easy,' says Jamie, 'but it's not. Bear in mind that *Acropora palmata*, the elkhorn coral, acts as a natural wave break. It's up there in the wave zone where it's really choppy. So you're in full scuba gear trying to do all this while you're being dragged backwards and forwards in the water, and being bashed by the rocks.' It takes a crew of around a dozen people to pull it off, but if something goes wrong, it's game over for another year owing to the crucial timing of the spawning.

Jamie's system changes all that. By tweaking the temperature, water chemistry, nutrient levels and duration and intensity of light, Jamie can now persuade his coral to spawn to order. Not only that; he can make his coral spawn more frequently than its wild counterparts, and so increase the amount of raw material for fertilisation. In the wild, coral spawn just once a year, but in the last twelve months, Jamie has organised four spawning events. He could arrange more, if he had the space and time. He can now predictably replicate the entire life cycle of a coral in a tank, and churn out new coral at a rate that far exceeds its wild equivalent. 'It's a game-changer,' says Keri.

Jamie's coral will never be exported because the logistics and regulatory hurdles are just too great, but Keri plans to use the knowledge she has gained to help restock the Florida Reef Tract. In Florida, one local species – the pillar coral – is now so endangered that there are only around 100 individuals left. They are struggling to reproduce naturally. 'We are up against a population that doesn't have the genetic diversity left to restore itself,' says Keri. 'Being able to cross the few remaining individuals by IVF could be vital to their survival.'

Jamie's technology means that Keri no longer has to face the challenge of trying to collect spawn from the wild, with all of its wave-battering uncertainty. Instead, she can take fragments of coral from the wild, grow them in tanks, use Jamie's method to make them spawn and then use IVF to create new individuals. When they are ready, they can be replanted onto the reef. At the Florida Aquarium's newly built Center for Conservation, four large tanks are being dedicated to coral restoration, where Keri and the team will be able to create the starting material needed to create genetically diverse coral colonies. 'It's the perfect place to take what we do to the next level,' says Keri.

Some Like It Hot

The project is exciting, but I wonder what is the point of restocking reefs if ocean temperatures continue to rise? With mass coral-bleaching events becoming more frequent, could we be approaching a time when there is nowhere wild left for coral to go, confining them to an indefinite future in captivity?

'Not at all,' says Keri. 'The Florida Reef Tract has been hit by high temperatures, land-based pollution and disease at an extremely rapid rate, yet there are pockets that still have healthy coral that look amazing.'

This is the key. There *are* colonies of coral that are stressed, sapped of colour and losing their grip on life, but sometimes, perversely, there are vibrant, healthy colonies growing right next door to them. In the midst of a coral graveyard, sometimes there is life. Corals have a delicate reputation, but some are incredibly hardy.

In Miami's Government Cut, for example, there are at least 16 species of coral. This polluted shipping channel is one of the busiest waterways in the United States. Over 4 million cruise passengers and 7.4 million tons of cargo pass through it every year. This isn't some pristine, tropical idyll; it's a dirty, man-made thoroughfare. It's a challenging environment for any sort of life. As the tide rises and falls, the inlet experiences wild shifts in water quality, level, temperature and salinity. Yet the corals living there are inexplicably thriving.

Equally bizarre are the corals bordering the coastlines of Saudi Arabia. Most corals bleach when the water temperature stays a degree or more above the local summer maximum for a few weeks. To put that in context, in many parts of the world, a sustained water temperature of 29°C is fatal. On one side of Saudi Arabia, however, the coral of the Persian Gulf routinely survives water temperatures that top 36°C, while on the other side, the coral of the Gulf of Aqaba, in the northern tip of the Red Sea, can survive temperatures of 33°C. So the question is; how do they do it? How do some corals manage to survive in these extreme environments? Why do some like it hot while others can't bear it?

In the context of global warming, the discovery of these hardy corals might seem like a godsend. Experiments suggest that despite rising temperatures, the corals in the Gulf of Aqaba are living well below their upper thermal limit. With their inherently tougher constitution, corals like this could still be around long after other, less stoical individuals have disappeared.

One obvious option is to transplant them. The Persian Gulf and the Great Barrier Reef, for example, share some of the same coral species, so maybe heat-tolerant coral from Western Asia could be relocated to the iconic Australian location? At face value, it makes sense, but conservationists have reservations. They worry about the spread of disease and point out that although the two sites are similar superficially, there are important differences. Subtle changes in salinity, upwellings, light, shade and various microscopic organisms could be enough to make any transplanted coral fail.

Instead, it's likely that these coral species will aid global conservation efforts indirectly, as scientists unravel the mechanics underpinning their hardiness. In time, they will pinpoint the idiosyncratic DNA sequences that endow these corals with their heat-tolerant nature. These sequences definitely exist and researchers will find them. These particular corals are the way they are because natural selection has been sculpting their genomes across hundreds of millennia. They teach us that, given time, corals can evolve to resist rising temperatures.

It is, however, a race to keep up. Human-driven climate change is in full swing. Global average temperatures have increased by about 1°C since pre-industrial times. Now experts agree that if they increase by a further 1°C, the resultant ocean warming, along with acidification, will lead to the widespread destruction of coral reef ecosystems over the next few decades. By 2050, there could be very few reefs left to save. Meanwhile, the corals are still evolving. They are just not evolving quickly enough. If only there was a way to speed up evolution.

Artificial Selection on Steroids

Ruth Gates is a Londoner who lives in Hawaii. By her own admission, she is fascinated by intimacy: the sort that exists between corals and their symbiotic partners. 'Corals are extraordinary exemplars of mutual symbiosis,' she says. They don't just form relationships with the algae that live inside their cells, they also have associations with bacteria, fungi, viruses and single-celled organisms called archaea. 'There's so much diversity in these relationships. It's just exquisite. It's the reason I started to study coral,' she says.

For most of her career, Ruth has been trying to work out if elements of these relationships predispose certain corals to cope better with stress. Then, a little while ago, she took a sabbatical.

'I began reviewing a lot of academic papers and every one said the same thing: *This work is directly relevant to the conservation and management of coral reefs*,' she says. But it wasn't. Not directly. 'There were all these good ideas out there but no one was making them happen.' Thousands of miles away on the east coast of Australia, Ruth's friend and fellow coral researcher, Madeleine van Oppen, shared the same frustration.

Then, in 2013, Paul G. Allen, co-founder of Microsoft, launched a competition called the Ocean Challenge. The challenge was for marine scientists to come up with new ideas to help mitigate the effects of climate change. He was looking for blue-sky thinking: practical, innovative solutions that could be applied directly to the ocean and make a real-world difference. The competition provided Ruth and Madeleine with an opportunity to crystallise two careers'-worth of nascent ideas into a 2,000-word essay, which they duly submitted. And their big idea?

The duo imagined being able to breed hardier corals through a variety of different methods including selective breeding, then using the fruits of their labour to restock the world's depleted reefs. In one report, they called it 'artificial selection on steroids'. Here was a plan with a direct conservation output. Critically, they weren't proposing to alter the coral's DNA via molecular methods, nor to make anything that wouldn't occur naturally in the wild, given enough time. Instead, they just wanted to give evolution a helping hand. So they called their idea 'assisted evolution'.

The judges of the Ocean Challenge loved their ideas, and the duo bagged first prize and a subsequent grant of US$4 million (£3.2 million). Now they are turning their dreams into reality. Ruth is Director at the Hawaii Institute of Marine Biology. It's situated on Moku o Lo'e (Coconut Island) in Kāne'ohe Bay off the island of Oahu. If you're familiar with the sixties American sitcom *Gilligan's Island*, then you'll recognise the island from the programme's opening sequence. It's a stunning place. Think palm trees, verdant foliage and tropical birds, set against a backdrop of crystal-clear waters that are bursting with coral. 'We have a living laboratory, right here in the bay,' says Ruth. 'We work 365 days a year on a living reef.'

In 2015, there was a massive bleaching event right on Ruth's doorstep. Thousands of corals in Kāne'ohe Bay shed their vibrantly coloured algae. It was heart-breaking to see, but the bleaching event also provided an opportunity. Fuelled by their new research grant, Ruth and her team pulled on their scuba gear and went out and tagged all the individuals that didn't bleach. Then the following summer, they went out to the reef again, collected eggs and sperm from these survivors, then used them for IVF in the lab. Studies have shown that the ability to tolerate warmer waters is heritable. Parents can pass this talent to their offspring. So the next step for Ruth and her team is to test their IVF 'babies' and see if they have inherited this valuable trait. If the experiments are successful, then the scientists envisage growing nurseries full of genetically diverse, thermally tolerant corals which could then be returned to the wild. Just as human IVF offers some childless couples a shot at parenthood, so too assisted evolution can offer some corals the same chance.

Meanwhile, Madeleine and her team at the Australian Institute of Marine Science in Queensland are doing parallel experiments using survivors from the 2016–17 bleaching event on the Great Barrier Reef. Although Kāne'ohe Bay and the Great Barrier Reef share a few species of coral, many are different, so it's important to establish the technology here too.

There's another potentially helpful difference between the corals in the two locations. In Kāne'ohe Bay, the corals are fairly prudish when it comes to sex: they only ever make offspring with members of their own species. But in Australia, where Madeleine works, some of the corals are more promiscuous. Sometimes the eggs and sperm from one species meet and meld with those from different species, producing hybrid offspring. Remember those urban Miami corals living in the busy, polluted shipping lane? The fused staghorn coral (*Acropora prolifera*) is a genetic mash-up of two other *Acropora* species, the staghorn coral (*Acropora cervicornis*) and the elkhorn coral (*Acropora palmata*). Part of the reason this coral is so hardy is precisely because of its mixed parentage. New genes can bring new life to tired genomes, and this is what happens with *Acropora prolifera*. This naturally occurring hybrid can withstand high temperatures and local pollution better than either of its parent species.

So now Madeleine and her team at the National Sea Simulator (SeaSim) in Cape Ferguson near Townsville, Australia, have been crossbreeding various local species of *Acropora* corals together in order to produce hybrids. 'We know that *Acropora* species occasionally hybridise in the wild over evolutionary timescales,' says Madeleine. 'We just wanted to increase the frequency of this naturally occurring process.' The experiments are promising. When the developing hybrids were exposed to warmer, more acidic water, they took it in their stride. 'We find that the hybrids are as fit or fitter than pure-bred corals,' she says. 'It suggests that by hybridising different species you might be able to restore lost genetic diversity and enhance the adaptive potential of these corals.'

Assisted evolution, in the guise of selective breeding, is reaping rewards. Just like Jamie, Ruth and Madeleine are realising the potential of coral IVF to help future-proof the world's reefs. By carefully choosing which individuals get to reproduce, the teams are guiding evolution and fostering the creation of new, resilient 'Super Coral'.

But these are not the only methods on the table. Madeleine is also creating 'Super Algae'.

Partner Swapping

There are hundreds of different types of zooxanthellae, the tiny, photosynthetic plant-like creatures that live inside the corals' cells, and they all have different qualities. Some, for example, are better at making nutrients; some are better at nurturing young coral; others are better at tolerating warmer waters. Critically, when these heat-tolerant algae are partnered with a coral, they give the coral some degree of protection, like a tiny, intra-cellular heat shield. Biologists have thought about inoculating corals with these thermally tolerant algae, but there is a problem.

Thermally tolerant zooxanthellae species are often poor photosynthesisers. They can withstand warming, but they don't make much energy for their hosts. Even though corals would benefit more in the long run from a partnership with one of these algae, they tend to reject them in favour of the ones that produce a more immediate reward. Who can blame them? If you had to choose between buying sun cream for next year's holiday or buying chocolate cake that you can eat now, would you think twice?

Madeleine, however, is hoping that coral can have their cake and sunblock too. For the last few years, she has been 'evolving' a culture of zooxanthellae in her lab. She took some of the algae that are decent food producers, then exposed them to elevated levels of temperature and acidity. As expected, many of the algae died, but the ones that survived were cultured across generations. Spontaneously occurring mutations that helped some algae cells to tolerate the stressful conditions were artificially selected for. Now, 80 generations later, the algae that have emerged are better able to tolerate the harsh conditions. 'When we put the algae back into the coral, there was a small benefit to the host,' says Madeleine. Swapping one strain of algae for another helped the coral to tolerate stressful conditions. 'The only problem,' says Madeleine, 'was that the effect we saw in the coral wasn't nearly as big as the one we saw *in vitro*.' So the game is still on. Madeleine and her team are trying to understand why the discrepancy exists. 'The good news, though, is that we did manage to evolve the algae and have the coral take it up,' she says.

Another conservation option is to manipulate some of the corals' other partnerships. In recent years, researchers have realised the importance of the microbiome, the myriad mass of microbial organisms that live in and on our bodies. Although a few do us harm, the vast majority of our gut-dwelling bacteria perform useful functions, including helping to digest food, break down toxins and fight off disease. Corals have their own microbiome, and are covered in a thick mucus layer that is teeming with bacteria. When the animals are stressed or starting to bleach, the ratio of 'good' to 'bad' bacteria begins to shift. Opportunistic and disease-causing bacteria predominate, leading researchers to wonder whether strategic interventions could restore microbial harmony.

There's certainly a precedent in other living systems. In humans, faecal microbiome transplants, or 'poo transplants' as they're known, are now accepted as an effective treatment of *Clostridium difficile* infections, while in agriculture, inoculating rice plants with microbes that come from plants growing in extreme environments can make the crop more resistant to drought and low temperatures. The idea then is to take the bacteria-containing mucus from healthy, thermally tolerant corals and use it to inoculate vulnerable or unhealthy corals, like a poo transplant for corals. 'These experiments are in progress,' says Ruth. 'If we can work out which bacteria are the important ones, then we may be able to make a freeze-dried version.' Probiotics for coral, no less.

When the Super Coral are supplied with their Super Algae and their Super Bacteria, the hope is that they will be able to survive the worst that we can throw at them, but there's one final tweak that is envisaged to give their assisted evolution a finishing veneer. Ruth and Madeleine want to send their coral to boot camp.

Coral researchers have known for over a decade that when a coral bleaches and then recovers, it is sometimes less likely to bleach the next time it experiences stress. 'It suggests they have some sort of intrinsic memory,' says Ruth. This protective response isn't governed by genetic mechanisms because the animal's DNA doesn't change. It's still the same animal and it still contains the same genes. What's different, however, is the way those genes are expressed. It's as if the coral has remembered that if it can switch certain genes on or off, it can protect itself from bleaching. The observation led Ruth, and her then PhD student Hollie Putnam, to wonder whether such epigenetic changes could be passed from parent coral to offspring. If an adult coral can remember how to survive a stressful event, can it pass this memory directly to its unborn offspring?

They decided to focus on a reef-building species called the cauliflower coral, which broods its young inside its body. The duo then exposed 'pregnant' corals to projections of global warming – warmer, more acidic waters – then let their larvae grow up, before exposing them to identical stressful conditions. They then compared them to a control group of larvae treated in exactly the same way, but whose parents were not stressed, and found that larvae from the first, experimental group survived and grew better. Not only that, but when the duo looked at markers of epigenetic change – tell-tale chemical molecules known as methyl groups – they found that the two groups were different. 'We think the existing molecular machinery is being re-tasked and passed on to the next generation,' says Ruth.

So Ruth and Madeleine have decided to be cruel to be kind. They've set up coral boot camps in their respective labs where they take the most thermally tolerant individuals they can find, then subject then to increasing levels of stress, all in the hope that they will pass this resilient nature on to their progeny. Ruth likens it to an elite-athlete training programme. 'We've gone out scouting for the talent, brought it into the gym, now we're running it on a treadmill.' The researchers are waiting to find out how these 'athletes' shape up.

A Brighter Future

Epigenetic changes are easily lost, so there's no guarantee that any inherited benefits will last. Nor will the training programme work for every coral species. Instead, it's important to embrace a wide range of possible fixes because coral are such a diverse group of animals. Some brood live young, but others are spawners. Some only associate with one type of algae, others with many. Sometimes species are shared between reefs and sometimes they're not. It's such a complicated and varied ecosystem that there is no 'one size fits all' approach that will work.

The coral reef research community is a bustling hubbub of enterprise and ingenuity. Faced with a challenge of epic and daunting proportions, they remain resolutely focused and optimistic. These are some of the most inspirational scientists I have ever had the pleasure to interview. Imagine being able to breed new strains and hybrids of resilient Super Coral in the lab, replete with Super Algae and Super Bacteria, which then go on to boot camp. When the corals grow up, they are replanted on the reef, but before they go, small fragments are broken off and returned to the lab. It's an insurance policy. The fragments are allowed to grow, then when they reach the appropriate size, conditions in the tank are tweaked to induce spawning, just as Jamie does in his south London lab. Clouds of eggs and sperm are released into the water at a prearranged time, enabling researchers to scoop them up and use them for IVF. The cycle of life begins again, providing scientists with new animals to study and plant out on the reef, and so it goes on. It's still evolution, but it's evolution that is guided by a well-meaning, informed human hand. This is conservation at its finest.

Conservationists imagine a global network of nurseries all dedicated to fostering future generations of coral. They picture scaling up production to reef-sized proportions. They imagine frozen repositories of coral cells: a biobank of precious genetic diversity and a back-up source of breeding material. Ruth imagines planes and satellites that can monitor the health of coral reefs from the skies. They will be fitted with sophisticated imaging technology that can detect the earliest signs of bleaching in individual corals, so that people can go out and intervene. She describes floating platforms that can be used to shade the reefs, and underwater pumps that stream cooling jets across their surface. In the future, this could all become possible.

This is not just about Ruth, Madeleine, Jamie and Keri, capable though they all are. This project is bigger than all of them. It is planet-sized. Thankfully, there are many other researchers, working in different parts of the world, on different reef systems, who all share the same goal – to future-proof and protect our precious coral reefs.

Now the biggest hurdle isn't making corals that can survive climate change. The biggest hurdle is alerting people to the urgency of the situation. If we delay creating new coral varieties until the reefs have practically gone, then so much genetic diversity will have been lost, that creating healthy, vibrant coral populations will become more difficult. And yet, Ruth tells me with a stoical expression, there are so many people out there who think that none of this is necessary. 'I think that people are paralysed by the scale of the problem,' she tells me. 'They think that if we keep doing what we're doing, and manage our oceans with protected marine areas, that everything will be OK.' It won't. These areas are not insulated from climate change. They cannot escape it.

Of course it would be better to draw a halt to climate change and hope that all of these problems will go away, but that's not going to happen. Even if global warming can be kept within 1.5°C above pre-industrial levels – which many now doubt is possible – shallow tropical seas will still warm enough to trigger frequent bleaching.

The corals, and the billions of life forms that depend on them, remain at risk.

In 2018, the head of the United Nations Environment Programme, Erik Solheim, warned that the battle to save the world's coral reefs is at 'make or break point'. He told the *Guardian* newspaper, 'Today I appeal to every single person on Earth to help us. We must replace the present culture of abuse with a culture of care.' A failure to act now, he cautioned, would bring about a major catastrophe.

Assisted evolution won't solve global warming. These projects are coral triage. They don't address the root cause of the problem, but they could buy the world's reefs some time while we rein in climate change. 'Assisted evolution may be able to increase the climate resilience of corals somewhat, but it will not be limitless,' cautions Madeleine. 'We need to deal with climate change!'

'At this point there's nothing to lose and everything to gain,' agrees Ruth. 'We need to create a movement of people that say "I can", rather than "I can't". I have this nascent feeling that there are enough interested people out there who care enough to make this happen. I'm not going to sit on my arse and do nothing.'

CHAPTER TEN

Love Island

Let me tell you all about the most enigmatic, charismatic and bizarre bird that I know. Imagine a large, obese budgie with an identity crisis. It has wings but can't fly. It's a parrot, but it's nocturnal. It makes a variety of noises, none of them bird-like. It 'purrs' like a cat, 'brays' like a donkey, 'wheezes' like an asthmatic, and 'booms' like the bass line of a house music anthem. Cat-sized, dark-clawed and blue-beaked, it sports verdant plumage. Its black, twinkly eyes are framed by enormous saucers of soft, yellow feathers that make it look like an avian Elton John, circa 1977. Say Kia Ora* to the kākāpō, one of the rarest and most intensively managed species on the planet.

The world's entire population of kākāpō live on a handful of islands scattered off the coast of mainland New Zealand. These are stunning, remote places with craggy, windswept hills and dense, green forests. The valleys are carpeted with soft green moss and the trails are lined with ferns and orchids. When author and wildlife enthusiast Douglas Adams† tracked them down

---

* This is Māori for hello.

† For those unfamiliar with Douglas Adams, why? Famous for penning *The Hitchhiker's Guide to the Galaxy*, he was one of the most talented, funny writers ever. Sadly missed.

for his 1989 radio documentary *Last Chance to See*, he claimed not just that the curious creatures had forgotten how to fly, but that they had forgotten they had forgotten how to fly. When alarmed, a kākāpō will sometimes run up a tree, then launch itself from a branch with all the élan of a world-class BASE jumper. With wings outstretched and gravity calling, what happens next has been described by some as 'gliding', by others as 'controlled free fall'. Adams, however, said the kākāpō flies 'like a brick'.

The species achieved infamy after Adams died, when *Last Chance to See* was remade as a 2009 BBC television series, and a sexually frustrated kākāpō called Sirocco ambushed Adam's friend, zoologist Mark Carwardine. The cocky male waddled out of the forest and climbed onto Carwardine's shoulders, then – wearing an expression that was equal parts determination and utter bliss – slapped him repeatedly round the ears with his frenzied, flapping wings. It was a literal head fuck. He rode the hapless presenter like a demented pair of sex-mad avian ear muffs. The incident prompted co-presenter, comedian Stephen Fry, to comment, 'Look, he's so happy ... You are being shagged by a rare parrot ... When you have the chick, I want you to call it Stephen.'

It was post-watershed TV gold, a moment of pure, innocent parrot porn that has since titillated more than 8.8 million viewers on YouTube. To put that into some sort of perspective, before Sirocco, YouTube's hottest parrot star was Snowball the Medium Sulphur-Crested Cockatoo who, at the time of writing, had racked up 6 million views simply by bobbing along to the Backstreet Boys. Sirocco took things to the next level. The debacle catapulted the kākāpō into wildlife documentary history, and Sirocco has become a comedy legend.

The kākāpō is worthy of our attention because its story is an evolutionary roller-coaster; an emotional journey with more highs and lows than The Big One at Blackpool.* This is a species that was driven to the brink of extinction by humans, and that now only survives because conservationists micro-manage its life. In an attempt to redirect the bird's future onto a firmer footing, biologists have experimented with everything from 'ejaculation helmets' and 'spermcopters' to 3D-printed eggs and high-tech genomics. Their work demonstrates the vital role of technology in shaping the future of today's endangered species. It's a tale of ingenuity and resourcefulness, of hope and determination, and of what can be achieved when people simply refuse to give up.

A Tribute to Mr Oizo

There was a time when there were hundreds of thousands of kākāpō spread all over New Zealand. They evolved in the splendid isolation of the country's primeval forests, an island paradise ruled by birds (see Chapter 5). Save for a few species of bats, there were no land-dwelling mammals, so the birds, reptiles and insects that lived there evolved to fill their ecological roles. Kākāpō, which feed on berries, seeds and other plant matter, assumed the role of a blue-beaked, tree-climbing rabbit.

When the first human settlers arrived, around 700 years ago, kākāpō were common. During the bird's

* The Big One at Blackpool Pleasure Beach is the UK's highest and fastest roller-coaster.

mating season, the settlers would have heard the deep, booming refrain of lovelorn males. Some have likened it to the sound of a heartbeat, others to the noise you get when you blow across the top of a bottle. Douglas Adams said it reminded him of the opening bars of Pink Floyd's 'Dark Side of the Moon', but I think it's a dead ringer for the bass line of Mr Oizo's 'Flat Eric'.*

Male kākāpō broadcast from specially constructed stages called 'bowls' which they build on high, rocky outcrops. They puff themselves up like feathered Pavarottis, and then boom, thousands of times a night, every night, from dusk till dawn, for months at a time. Interspersed with the occasional asthmatic-sounding 'ching', this low-frequency serenade carries across the valleys and can sometimes be heard several kilometres away. Females listen out for this curious song and then gain entry to the bizarre music concert via a radial network of tracks that are prepared and meticulously maintained by the feathery artist himself. This is the only parrot to operate a lek breeding system. This is basically a big showing-off competition where individual males vie to sing the sexiest song, then the 'winner' gets to mate. It's like an avian X Factor.

With kākāpō providing the booming soundtrack to the Maōris' everyday lives, the birds became etched into Maōri folklore. The early Polynesian settlers noted that kākāpō breeding tends to coincide with the erratic fruiting of the bird's favourite food plant, the rimu tree,

---

* The inexplicably popular 1999 house music hit. It spawned a TV advert featuring a yellow, wide-mouthed, dead-eyed puppet in a car, tapping along to the beat. It never ended.

so they believed that the birds could tell the future. They couldn't, of course, which explains how the Māori were eventually plucking them, skinning them and cooking up a storm. The bird's natural history came to an end as its fate became inextricably interwoven with our own.

Hopelessly naive around predators, kākāpō were easy pickings. The Māori caught them using snares, traps and Polynesian dogs. Then they used their feathers to make cloaks* and their meat to make dinner.

By the time Europeans arrived, in the early 1800s, the kākāpō population had taken a hit and its range had diminished. The birds could now only be found in the central North Island and forested parts of the South Island. Where the Polynesians introduced just two non-native species; the dog and the Polynesian rat, Europeans introduced all manner of predatory enemies. First came two more species of rat; the big, black ship's rat and the even bigger, brown Norwegian rat, which feasted on the kākāpō's eggs and chicks. To catch the rats, whalers stocked their ships with cats, but when the ships weighed anchor, the cats ran away and started eating the endemic ground-dwelling birds. The kākāpō was particularly vulnerable because its strange scent – an earthy, honey-like smell – made it easy for predators to find. Next, the settlers brought pigs, deer, goats and sheep, which trampled the ground and competed with the kākāpō for food. Hunters brought rabbits, which

---

* Each cloak incorporated more than 10,000 feathers, making kākāpō cloaks the ultimate status symbol of the time. Today, the phrase 'you have a kākāpō cape and you still complain of the cold' is like telling someone they moan too much because their diamond shoes are too tight.

ran amok and bred 'like rabbits', so they next imported ferrets, stoats and weasels to catch them, and these also ate the native birds.

By the end of the nineteenth century, the kākāpō left on mainland New Zealand were mainly confined to the steep slopes of Fiordland, a mountainous, craggy region towards the base of the South Island. The government decided to launch its first official rescue mission and under the guidance of an Irish-born conservationist called Richard Henry, hundreds of kākāpō were captured and moved from the mainland to a smaller, predator-free island.

Although the practice is commonplace today, at the time it was remarkably forward-thinking. New Zealand is made up of around 600 islands, so it made sense to relocate the birds to one of the country's more remote isles that was free of non-native species. Resolution Island (Tau Moana), the country's seventh largest landmass, was set aside as a dedicated nature reserve, but sadly, it wasn't stoat-proof. The animals swam there from the mainland and eight years later, all the kākāpō were dead. Richard Henry was heartbroken, and although he remained on Resolution Island for a while, he eventually retired. As for the kākāpō, few people heard or saw them after that. No one actively cared for them and the conservation programme folded.

By the middle of the twentieth century, most people thought the kākāpō was extinct, but in the early 1950s, the newly formed New Zealand Wildlife Service decided to find out definitively. They launched more than 60 expeditions to look for the kākāpō and eventually they were rewarded when tracker dogs discovered 18 birds hiding in the remote wilderness of Fiordland. Spirits were high but the celebrations were short-lived. When the birds were caught and inspected, there was a problem. All of the kākāpō were male. The New Zealanders didn't give up and kept on searching, and in 1977, they discovered a second, much bigger population of kākāpō living on Stewart Island (Rakiura), the country's third largest landmass.

Critically, this group of 200 birds included females, raising hopes for the species' survival, but although there were no stoats or weasels on Stewart Island, there were feral cats, which were rapidly eating the kākāpō. The decision was taken to relocate the kākāpō, but instead of putting all their eggs in one basket – or birds on one island – this time they decided to spread the risk. All the Stewart Island birds and five of the male Fiordland birds were split between a couple of offshore islands, but sadly, their decline continued. Breeding was erratic and the species' numbers reached an all-time low. In 1995, there were just 51 birds left alive.

It was the kākāpō's lowest ebb. The species was fast heading the way of the dodo. It became obvious that the birds needed more support, and that relocating them wasn't enough. So in 1996, the Kākāpō Recovery Programme was born. It was the beginning of one of the most intensive and inventive conservation initiatives that the world has ever seen.

Love Island

The Kākāpō Recovery Programme is a real-world version of the hit TV show *Love Island*, where sexy singles are sent to live on a distant island paradise. Just like their television counterparts, the kākāpō have all been deliberately placed in this location by humans, who then manage their environment and monitor their behaviour. The birds have plentiful food, souped-up accommodation and a ready supply of genetically compatible partners. The islands are laced with hidden cameras and surveillance equipment. Instead of microphones and wires, the birds wear backpacks and transmitters, so the humans can eavesdrop on their most intimate moments. There are libertines and Lotharios, loners and those unlucky in love. In addition, kākāpō are naturally promiscuous, so there's no shortage of partner-swapping and plot twists. Sound familiar? Just like the TV show, the preening kākāpō must couple up in order to survive, but there's more than just ratings at stake here. The survival of the entire species hinges on the outcome of these carefully monitored trysts.

Andrew Digby is Science Advisor for this uniquely avian Love Island. It's his job to monitor and research the birds, understand their needs and dream up strategies to ensure their survival. Now he stands on a hilltop surveying the scene. At his feet, the green, scrub-covered slopes drop away to merge with the distant grey waters of the Southern Ocean. The rain has finally eased. For the last few weeks, it's been hammering down, making the tracks that lead to the summit muddy and treacherous. Now the clouds have parted, the sand flies have made their presence known, and Codfish Island (Whenua Hou) is alive with the sound of birdsong.

In his khaki jumper, shorts and walking gaiters, Andrew stands straight with his right arm aloft, clutching what seems to be a rickety, old-fashioned TV aerial. Striking the pose, he looks like an Antipodean Statue of Liberty. A British-born former astronomer, who spent time with NASA looking for planets in distant solar systems, Andrew is now looking for an equally elusive target: kākāpō.

Kākāpō are reclusive birds. 'Kākāpō' is a Maōri word: 'kaka' means parrot and 'po' means night. During the day, when they are sleeping, they are almost impossible to find. Their mossy-coloured feathers help them to blend into their background and effectively disappear, but the signal transmitting from their backpacks can be used to locate them.

Uniquely for any wild species, every adult member of the species wears a backpack. The lightweight accessory fits snugly around the bird's wings and carries a transmitter that emits a unique radio signal that can be used to identify each individual bird. Today Andrew is looking for a very particular male: Blades.

Blades is an enigma. One of the original cohort that was discovered on Stewart Island, he's incredibly popular with the females. 'He's one of our most prolific breeders,' says Andrew. 'We've no idea why.' Between 2002 and 2016, Blades fathered 20 kākāpō. To put that in context, in 2016, Blades was responsible for 13 per cent of the world's entire kākāpō population. The team were worried that this single bird was beginning to monopolise the gene pool, so in 2017, the decision was taken to relocate him away from Codfish Island, where there are more females, to Little Barrier Island (Hauturu) in New Zealand's far north, where there are far fewer females. 'We're banishing him,' quips Andrew.

Catching him, however, is no mean feat. The 'TV aerial' that Andrew is holding is a radio telemetry unit that has been tuned to detect Blades' signature frequency, so now 'all' Andrew has to do is head to a high point, hold the receiver aloft and then hone in on the signal. This is easier said than done on an island that is studded with hills and valleys, trees and dense undergrowth. It involves crashing around the thorny scrub in ever-decreasing circles until the elusive bird is finally located.

When Blades is eventually found, in the shade of a rata tree, he succumbs to his fate without resistance. Andrew plucks the bird from the greenery and carefully places him inside a pet carrier. Then he is taken to the beach, helicoptered to Invercargill Airport on the South Island, and then flown by Air New Zealand plane to Auckland on the North Island. He rides in the cabin alongside the regular human passengers and three other kākāpō who are also making the trip. Then from Auckland, it's a short hop to Little Barrier Island. Twenty-four hours after his ordeal began, Blades and his compatriots are no worse for wear, so the kākāpō are released into their new home. Blades looks mildly nonplussed as he eyes his new surroundings, but the confusion doesn't last long. The jet-setting Casanova shuffles out of his pet carrier and as he bumbles off into the undergrowth, a new episode of *Kākāpō Love Island* has begun.

A Feathery Roll Call

Kākāpō are so charismatic that it's hard not to anthropomorphise them. It's a tendency that is further fuelled by the team's tradition of naming each and every one. Blades was named after Bill 'Blades' Black, a legendary helicopter pilot who transported the team, sometimes alarming his passengers by flying his helicopter with his knees while he lit his pipe. Sirocco was named after the hot, dry North African wind. There's Rangi, named after the Māori word for 'sky'; Richard Henry, named after the nineteenth-century conservationist who tried to save the kākāpō; and Bluster Murphy, who hatched on a blustery night and later lost two toes when he was attacked by another male kākāpō. Joanne Paul-Murphy was the vet who saved his life.

When I first spoke to Andrew, back in 2016, there were 154 kākāpō. Thanks to the team's hard work, the population had more than tripled from its lowest point back in 1995. This is a heart-warming statistic. Twenty-five years ago, the kākāpō were careering towards extinction, but now their evolutionary journey has been steered onto a safer path. Progress has been slow and steady, but sometimes it feels frustrating that things aren't moving more quickly.

Perhaps the biggest hindrance has been the kākāpō itself. It's a very unhelpful bird that seems to do little to further its own existence. When species evolve in predator-free environments there is no urgency to start a family (see Chapter 8), so kākāpō take at least five years to reach sexual maturity. Even then, they often don't rush into things and have very specific requirements. Adults breed erratically, every two to four years, and only when there is a bumper crop of berries on the rimu trees. With mating over, the male's work is done. Male kākāpō are deadbeat dads who offer nothing in the way of childcare, so the females are left to prepare their nests and raise their offspring by themselves.

The nests are pretty basic affairs: empty spaces found inside hollow logs and rotten stumps, or under knotty tree roots. Sometimes, they let in rain and predators, and are not the sort of secure nurseries that conservationists would wish for one of the world's most endangered birds. Females then lay two to four eggs (more would be nice), which they then incubate for around 28 days. This is when the pressure of being a single parent really takes its toll. The females have to vacate their nests at night in order to find food. This leaves the eggs – and the chicks when they hatch – unguarded and vulnerable. To top it all, today's kākāpō are suffering from a fertility crisis. Around half of all eggs that are laid are infertile and it's not uncommon for chicks to die after they have hatched.

So now science finds itself at the heart of their epic recovery. 'Our work is very technology focused,' says Andrew. 'We've tried so many different things. It really pushes the boundaries in terms of what's possible for conservation.'

Breeding seasons are hotly anticipated. The team can predict their onset by studying temperature patterns, and by counting the number of fruit that are developing on the rimu trees. Rimu fruit are slow to develop, so the team knows that if more than 8 per cent of the trees' tips have fruit, then the following year the kākāpō will breed when the berries are ripe. It's their klaxon call to get busy.

The usually 10-strong team recruits a mini-army of additional helpers who volunteer to stay on the islands and work as 'nest minders'. Kākāpō sometimes re-use old nests, so any dilapidated nests are renovated. If they're broken, they're fixed. If they leak, they're waterproofed.

If they're beyond repair, then welcoming nest boxes are placed next door. Because the nests can sometimes be deep and inaccessible, they're also fitted with artificial hatches so the nest-minders can access their contents. Surveillance equipment is then installed: an infrared camera inside the nest and an external device called a snark.* These boxes log and relay transmitter signals, and are placed immediately outside the nest so that when a female leaves to go foraging, the device picks up her signal and alerts the nest-minders.

Smart hoppers, stocked with additional food, are dotted around the island. The team has realised that mothers who receive additional food spend less time away from the nest and are more likely to raise a healthy brood, so these head-height boxes are programmed to pop their lids only for those birds whose signals they have been programmed to recognise. Smarter still, the kākāpō must stand on a set of weighing scales to reach the hopper, so the team can also weigh them and adjust their feeding regime as necessary.

Breeding season usually begins in January or February as females respond to the males' booming operatics. The 'action' occurs off camera under the cover of night, but during breeding season, little in the kākāpō's lives goes unrecorded. The backpacks of male kākāpō contain a modified transmitter called a 'Check Mate' which can detect the frenzied flappings of a mating male. Cleverly, the Check Mate also contains a receiver which detects

---

* Snarks are named after the Lewis Carroll long poem, *The Hunting of the Snark*, which, fittingly, is about an enigmatic bird that was impossible to find.

the female's transmitter, so the team can pinpoint which female was involved. And if that's not enough to dampen the male's ardour, the encounter is automatically awarded a score. 'The device also monitors how long the interaction lasts for,' says Andrew. 'It even gives us a measure of the quality of mating.'

Females then nest around a week after mating. The team knows when this happens because a specially adapted transmitter in the females' backpacks, called an Egg Timer, recognises when the females have stopped moving around and settled down to lay their eggs.

Nest-minders are invited to come and *not* sleep in specially erected tents, which receive a live video feed from the nearby nest. When the female goes out foraging, the snark triggers a doorbell noise inside the tent, which alerts the campers. It's their signal to wriggle out of their sleeping bags, pull on their boots and go and check on the clutch.

Early on, while the developing chicks are still in their shells, the eggs are removed and 'candled'. The fragile structures are held in front of a light source so that the contents can be seen in silhouette. It enables the team to assess whether or not the eggs are fertile. Then, after the healthy eggs hatch, the team make regular visits to weigh and check the chicks. If a chick is in trouble, then it may be physically removed. Early on, the team realised they could help new mothers by hand-rearing chicks that were sickly or slow to gain weight, but it's a cautionary tale …

Sirocco's Story

Sirocco was born on Codfish Island in 1997. 'Newly hatched kākāpō are really fluffy little things,' says Daryl

Eason, who is the Technical Advisor for the Kākāpō Recovery Programme. 'They're covered all over in short, white down. We'd go into the nest and weigh Sirocco and his brother every third night while the mum was away foraging.' All went well for a while, but then Sirocco's growth slowed and he developed a wheeze. 'He sounded like he was asthmatic,' says Daryl. So the decision was taken to remove him from the nest and hand-rear him.

That night, an indignant Sirocco was transferred into a bucket and carried to the rangers' hut where they converted one of the bunk rooms into a makeshift nursery. 'I put Sirocco into a big red tub and began to feed him immediately,' says Daryl. In the wild, mothers feed their begging babies by regurgitating food from their crops, but Sirocco was too weak to beg, so Daryl fed him through a tube. 'He took to it really well,' says Daryl. 'He became very keen to feed. He'd flap his wings and grunt like a little pig.' Sirocco turned a corner. His breathing problems cleared up and he began to put on weight.

When he was three months old, Sirocco was moved to an enclosure outside the hut. The plan was not to keep him captive, but to get him used to being handled so that if he needed to be captured in the future, it would be easier to do. 'It was wonderful spending time with him,' recalls Daryl. 'He'd climb all over you and play-fight with your hand, just like a cat. It was the first time that any of us had been in such close contact with a kākāpō.' But it led to problems. Sirocco became imprinted on humans, and to this day, he prefers the company of people to the company of kākāpō. This explains his attraction towards Mark Carwardine's head.

To that end, the Mark Carwardine event is far from an isolated incident, and Sirocco's sexual tastes are about as liberal as they come. When he's in the mood, he'll try to mate with just about anything human-related, including heads, shoes and jumpers. When he was released from his enclosure, he set up his bowl close to the hut so he could sing to the humans inside. One can only suppose that his plan was to lure them from the hut and it worked, not because they were attracted to him, but because they had to walk past his bowl to get to the outside lavatory. 'If there were people walking by, he'd rush out, jump on the back of their legs and try to climb up them to get to their heads. If you didn't know how to deal with him, it could be pretty intimidating,' Daryl recalls. 'Eventually we had to put plywood fences around the track so people could go to the toilet in peace.'

In the end, the avian Michael Douglas needed 'sex therapy' to curb his unusual impulses. A professional animal trainer called Barbara Heidenreich was called in, who managed to 'cure' Sirocco through the use of positive reinforcement and macadamia nuts. He learned to channel his ardour away from humans and towards an inanimate object instead. When Sirocco lived near the hut on Codfish Island, he used to steal the soft, plastic 'Croc' shoes that belonged to the rangers, so because he already had an interest in Crocs, it was decided to make these the new focus of his amorous intentions.

Now, he no longer mates with people's heads, and the shoes have been re-named 'Krokapos'. If Mark Carwardine were to visit him again, I suspect he'd be slightly disappointed to find that Sirocco no longer fancies him. There has to be a certain cachet from being shagged by one of the world's rarest birds … and a crushing sense of disappointment when you find out that you've been dumped for an ugly shoe.

The Kākāpō Recovery team has learned from Sirocco. His story illustrates both the promise and the perils of hand-rearing. The practice is still used, but human interference is kept to a minimum. As a result, more than 65 sickly chicks have been successfully nurtured to adulthood without any imprinting issues, and Daryl, who has cared for them all, is affectionately referred to as their 'surrogate mum'. But his talents don't stop there. He also manages a fostering service.

Over the years, the team has realised that kākāpō mothers are tolerant birds who will readily raise the offspring of others. Now, if a nest is in danger, or if a mother has too many mouths to feed, her youngsters might be hand-reared by humans, or better still, they might be given to a foster bird to look after.

Kuia, for example, is a very special bird. Of the 18 kākāpō discovered in Fiordland, only one bred and passed on his genes. His name was Richard Henry. Kuia is his daughter, so she carries the Fiordland genes. This makes her very precious indeed, as the Fiordland birds carry unique genetic variants that are not found in the Stewart Island birds. In 2016, she bred for the first time, but chose a terrible place to nest. It was close to a seal colony and plagued by sand flies, causing Kuia to become flustered and break one of her eggs. So the team destroyed her nest and gave the rest of her clutch to females in safer nests whose own infertile eggs had failed to hatch. The foster mums raised the chicks as their own, while

Kuia re-mated and moved into a better nest where she raised a second brood of her own.

Now the 2019 breeding season is in full swing. By studying Kuia and other birds, the team has learned that if a mother's eggs are removed before they hatch, it spurs the female to re-mate and lay a second clutch. So for the first time ever, Daryl, Andrew and the team have decided to steal all of the first clutch and hand-rear them, in the hope that the females will produce a second brood. 'It's quite a gamble,' admits Andrew. 'We've done it for 12 nests in the past. Now we're going to do it for many more.' It's hoped the strategy will double the amount of kākāpō that are produced this season, which could make a significant difference to the much-beleaguered bird.

In addition, they'll be performing an altogether more delicate manoeuvre: artificial insemination. It might sound intrusive, but there are multiple reasons to think that this is a very good idea. First, artificial insemination (AI) has been used to swell the numbers of other at-risk birds including Spix's macaw, the white-naped crane and the Magellanic penguin. Second, if the procedure works, then it will give a male who is fertile yet unlucky in love the opportunity to pass on his genes. This boosts overall levels of genetic diversity, helping to create a more robust and resilient population of birds. Third, kākāpō are polygamous, meaning both sexes mate multiple times with different birds. Research from the team suggests that if a female mates with multiple partners it may increase her fertility. So AI doesn't just increase genetic diversity, it also increase a female's chances of becoming a mum. Now, if a tryst is deemed unsuitable – if the birds are related, too young, or already over-represented in the gene pool – AI is considered. Which begs the question, how does one go about avian AI?

Back in the nineties, before Andrew joined the team, conservationists tested some pretty wild ways of collecting kākāpō semen. There was Chloe the remote-controlled car, a toy jeep with a kākāpō puppet glued to its boot. With its seductively splayed wings and prominent rear end, the hope was that amorous males would jump on board and 'go for a ride', but the birds weren't fooled. Chloe was rejected and her mechanical rear was consigned to the scrap heap. So next they tried the 'ejaculation helmet', a glorified swimming cap that was covered in artificial cloacas.* Not for the faint of heart, the helmet was worn by members of the team who then loitered in the undergrowth trying to look sexy. Yet again, the hope was that frisky males would mate with the device – much as Sirocco mated with Mark Carwardine's head – and leave their seed in the helmet's pitted plastic receptacles, but the birds weren't interested. Now the ejaculation helmet has also been consigned to the history books and the team have adopted a more subtle and successful approach.

Today, if AI is to be performed, radio telemetry is used to track down a male, who is then held calmly and massaged gently around his spine, abdomen and pubic bones. The bird does the rest. The waiting female is then flipped over so her belly is skywards and the sperm is inserted into her oviduct via a small plastic catheter.

---

* A cloaca is the 'hole' through which female birds poo, wee, have sex and lay eggs.

Insemination is a speedy affair and within a couple of minutes, the female is free to go.

It sounds simple, but as ever, there are complications. 'One problem is that sometimes the male we want is on one side of the island, and the female we want is on the other,' says Andrew. It can take hours to deliver the precious semen on foot, so in 2019, the team tested their latest innovation, the 'spermcopter'. Andrew happens to be a dab hand at flying remote-controlled drones, so now when the team need to deliver semen from one side of the island to the other they tether it to a drone and launch it into the air. 'Now a journey that would take an hour on foot, takes around four minutes by drone,' he says.

The biggest issue, of course, is choosing which birds should donate and receive semen. It's important to avoid incestuous pairings as they can lead to inbreeding, but it's not always obvious how different birds are related. Because females mate with multiple males and because males play no role in childcare, paternity is never obvious. For these reasons, scientists are turning to DNA.

Cracking the Kākāpō Code

In 2003, Bruce Robertson from New Zealand's Otago University developed a 'relatedness' test for kākāpō based on small snippets of their DNA. The test is based on DNA fingerprinting, a genetic method that was originally developed to identify people. It's an incredibly powerful technique that has been used to prove guilt and innocence in criminal cases, resolve immigration disputes and clarify paternity, but in kākāpō its use was limited. Although the test can determine if two kākāpō are related, it can't define the exact nature of that relationship. It can't tell the difference between siblings, cousins or parent-offspring pairs; crucial information when you're trying to match-make one of the world's rarest birds.

Andrew realised that if he had the full genetic sequence or 'genome' of every single bird, rather than the fragments they used for DNA fingerprinting, then the foggy kākāpō family tree would become resolved with crystal clarity. Detailed genomic information could be used to help guide the kākāpō's conservation.

A similar resource already exists for another avian species: a big, bald, beaked scavenger called the California condor. Thirty years ago the species was almost extinct after being poisoned by the lead shot inside the carrion that it ate. Then, in 1987, conservationists decided to take all the remaining wild birds into captivity and establish a breeding colony, but the endeavour was blighted by the emergence of a genetically determined disease that caused a lethal form of dwarfism. So they sequenced the genomes of 36 California condors – a significant proportion of the total population – and used this information to create a detailed genetic map. The map helped them to resolve the bird's family tree and to identify and exclude disease-carriers from the breeding programme. Now the programme goes from strength to strength, and many of the captive-bred birds have been released into the wild. Today, there are 290 California condors riding on the thermals above California, Utah and Mexico.

Now imagine something similar for kākāpō: an avian Match.com where virtual pairings are given a compatibility score based on their genome sequences and life history. High-scoring couples could then be matched in real life by placing them on the same breeding island and/or by the use of AI. Meanwhile, low-scoring couples that are closely related or have fertility problems would be geographically separated. DNA data could be used to inform practical conservation decisions. 'It could be a game changer for kākāpō conservation,' says Andrew.

Beyond that, the genomes could be used to address other fundamental issues. If the original kākāpō DNA test is like working with shreds of pages torn from a book, then having the full genomes of all the birds would be like having an entire library to browse. 'It would be a massive resource to have,' says Andrew. 'There's so much about the kākāpō that we don't know. It would give us so much more insight into the problems that we have.'

Top of the list is to find out whether there's a genetic basis to the high rates of infertility experienced by the species, and why some individuals are more prone to disease. After that, there are more basic conundrums to address, such as how long do these birds live for? No one knows hold old the founding birds were when they were discovered back in the seventies, so although they must be more than 40 years old now, it's impossible to predict how much of their lives remains. 'We'd like to know this,' says Andrew. 'We think they can live for up to 90 years, but we're just not sure.'

Another question relates to the birds' evolutionary history. Studies suggest that kākāpō belong to the so-called 'basal clade' of parrots, the group that was the first to split from the common ancestor of all of today's parrots. Two other New Zealand birds, the kea and the kākā, also belong to this clade, but the kākāpō is more distinct and is the only species that has its own specific genus (*Strigops*) – meaning it could be the most ancient parrot species alive today. Unravelling the kākāpō genome could help to explain why some birds are smarter than others and why parrots are so good at mimicking people.

Scientists have decoded the genomes of many different species, including pelicans, electric eels and wallabies, but when they do this they tend to decode the DNA of a single reference specimen. The Genome 10K project, for example, aims to sequence the genome of at least one individual from each of the 66,000 or so vertebrate species, while the B10K project seeks to decode the genome of at least one individual from all of the 10,000 bird species. Andrew was suggesting something very different. He dreamed of decoding the genomes of as many different kākāpō as possible, both alive and dead.

Fortunately for Andrew, the world of academia is full of people with very specific, niche interests, so he was not alone in his yearning to delve deeper into the kākāpō's DNA. At Duke University, geneticist Jason Howard became interested in the kākāpō after he read a picture book called *Kākāpō Rescue* with his young daughter. Jason, who was working on the B10K project, says, 'I was reading it aloud and it was obvious. I thought, we really need to sequence this bird's genome.' So he teamed up with Bruce Robertson and the Kākāpō Recovery team, and was given a blood sample from a kākāpō called Jane.

Jane was chosen because she had fertility problems, so the scientists thought it would be a good way for her to contribute to the future of the species. Then, after he'd extracted and sequenced the DNA from her blood, Jason was approached by Andrew. Jason knew that once you have the genome of one individual from a species, it's then comparatively easy to decode the genomes of additional members. 'The hard part is finding the funding to do it,' he says.

Andrew had an idea. He knew of a freshly formed outfit called the Genetic Rescue Foundation. It was set up by a computer scientist, David Iorns, who wanted to help make genetics-based conservation projects happen. Within three days of the organisation's website going live, Andrew was on the phone. David, who hails from New Zealand and knew of the kākāpō, was smitten with the idea of sequencing the whole species' genome. An agreement was established, a crowd-funding webpage was set up and a new project was born. Christened 'Kākāpō-125' after the number of kākāpō that were alive when the plan was conceived, Andrew's dream began to morph into reality.

Since then, Andrew and the team have collected blood samples from adult kākāpō during routine health checks, and a couple of years on, the project is all but complete. It follows in the footsteps of similar work on the Spix's macaw, another desperately endangered parrot, which recently became the first species where every individual has had its genome decoded. Now 169 kākāpō have had their genomes sequenced and the data is already starting to resolve some mysteries.

In 2018, Swedish researchers compared the genomes of modern kākāpō with older historical specimens, and found a marked decline in genetic diversity over the last

200 years. This suggests that early Polynesian settlers had little impact on the kākāpō, and that much of the species' decline is attributable to the later arrival of European settlers and the invasive species that they brought.

It also sheds light on the origins of the Stewart Island population that was discovered in the seventies. Originally, people thought that Polynesians or Europeans had introduced the kākāpō to the island in the nineteenth century, but the study suggests there have been kākāpō on Stewart Island for much of the last few thousand years. If both scenarios are true – which is entirely possible – then there were already kākāpō living on Stewart Island when the mainland birds were imported. This is good news because it means the birds descended from Stewart Island stock could be more genetically diverse than was originally thought.

Baby Boom

In 2018, Santa Claus visited New Zealand right on cue. The first eggs of the 2018–19 breeding season appeared in a nest on Codfish Island on Christmas Eve. It was the best present that the Kākāpō Recovery team could have hoped for.

Perhaps in response to the phenomenal burst of rimu berries that had erupted on the trees, breeding began early. The Check Mate transmitters sprang into life in mid-December, as the booms of hopeful males reverberated through the valleys. Now, three months later, Andrew and the team are reaping the rewards of the most successful breeding season that the Programme has ever known.

There's a whiteboard on the fridge door in the rangers' hut on Anchor Island (Puke Nui), one of the kākāpō breeding hotspots. Scribbled on it, in thick black marker pen, is a list of the 50 adult breeding females. There's Pearl and Pounamu, Solstice and Sue. Kuia is among them too, with her precious Fiordland genes. Next to each name, there is a little row of eggs that denotes the progress of each bird. Some have smiley faces because the eggs they symbolise are fertile. Some have smiley faces, wings and feet, because the eggs they represent have hatched. Ovals marked with a straight line or a cross stand for eggs that were infertile or embryos that died inside their shells, while parallel lines indicate a female that has re-nested and produced a second brood. At the end of March 2019, there were 132 infertile eggs, but there were also 72 chicks and 117 fertile eggs waiting to hatch.

'It's been a phenomenal season,' says Andrew. 'So far, we've had 249 eggs laid.' To put that in perspective, in all of the breeding seasons between 1981 and 2017, a total of 410 eggs were laid. Today's tally of 249 is 60 per cent of that number, all laid in just a few short months. Finally the pace is picking up, and the team is reaping the rewards of decades of hard work.

The spermcopter has been out and the team has performed more than a dozen artificial inseminations. Now they are waiting to see if the eggs that result are fertile. The decision to hand-rear the first round of eggs and force the females to re-mate has paid off. Thirty females re-nested and produced a second clutch of eggs, seriously swelling the number of developing chicks. Kuia's first clutch was all infertile, but the three eggs in her second batch are now starting to hatch. Meanwhile her brother, Sinbad, who is also of Fiordland descent, has mated and now the team must wait for the results of DNA tests to work out the paternity of this season's chicks. If Sinbad has fathered chicks, this will be excellent news as his unique Fiordland genes begin to trickle through the population. Daryl, the human surrogate mum, has been busy, as record numbers of chicks are being hand-reared and cross-fostered. Meanwhile, Andrew can't stop tinkering with technology. He struck up an alliance with Weta Workshops, the special-effects company that made the creatures and costumes for *The Lord of the Rings*. Now the company has produced arguably its finest work to date – a 3D computer model of a real kākāpō egg. Using this as a template, Andrew has been able to 3D print a whole series of kākāpō eggs that are exact replicas of the real thing. Now, if the team needs to place a dummy egg in a nest, for example if they want to prepare a female for an impending foster chick, then the eggs they have are of Middle Earth quality.

Meanwhile, it's now 20 years since Sirocco was hand-raised by Daryl, and 10 years since 'that incident' with Mark Carwardine. Sirocco is a reformed character. When the team realised that he was unlikely ever to mate with another kākāpō, they found a different role for him. They decided to put his high-profile chutzpah to good use and in 2010, the then Prime Minister for New Zealand, John Key, made Sirocco the country's 'Official Spokesbird for Conservation'. Said Key at the time: 'He's a very media savvy bird, he's got a worldwide fan base – they hang on every squawk that comes out of his beak. He'll be a great official Spokesbird and a great ambassador for New Zealand.'

He really is. Now Sirocco divides his time between 'chilling' on his remote island home and touring the country, meeting adoring human fans and raising the profile of New Zealand's endangered species. In an ironic twist, he's even learned how to fly (first class by plane), and has acquired more than 20,000 followers on Twitter ... which is also ironic, given he can 'boom', 'bark', 'skraaarrk' and make all manner of weird noises, but he can't actually 'tweet'.

The kākāpō is a big, beautiful, barking species that smells like a violin case and lives in one of the most unspoiled, isolated places on Earth. If you were to parachute in, outside of breeding season, you'd no doubt marvel at their unfettered and seemingly wild life, but the wild you would see would be a carefully managed illusion. This is a species whose evolution *was* heading towards oblivion, until humans stepped in and decided to turn things around. Now the kākāpō can only live on their Love Islands because people actively manage them. There are no predators, because humans keep them out. Conservationists modify their nests, tinker with their diets and organise their procreation. They use the latest technology, and as the genome project nears completion, the Kākāpō Recovery team is now moving from managing the birds' breeding to micro-managing it. With each bird's DNA decoded and stored in digital form, decisions about the species' future will be increasingly informed by computer algorithm, adding a new layer of sophistication to the Kākāpō Recovery Programme. So I wonder, will there ever be a time when Andrew and the team can step back and stop managing the species? Will the birds ever be able to shrug off their backpacks and fend for themselves? Will they ever regain control of their evolutionary journey?

'That's our long-term plan,' says Andrew. 'The birds are very heavily conservation-dependent at the moment, which is an issue, but I feel optimistic. It all boils down to pest control and having predator-free areas that are suitable for kākāpō.' If New Zealand's Department of Conservation makes good on its promise, and makes the country predator-free by 2050, then the kākāpō will be just one of many species to benefit. In the meantime, the kākāpō team walks the tricky tightrope between heavily interventionist conservation and keeping the kākāpō population as wild as possible.

Now the birds are split between a handful of islands that are more or less managed. Blades, remember, was banished to Little Barrier Island because he was a super stud. Now he lives there with a small group of renegade kākāpō who still wear backpacks, but are pretty much left to their own devices. 'We still rely on the technology to monitor them, but it's a much more hands-off approach,' says Andrew. The team is experimenting with managing the birds less, while at the same time scoping out new habitats for their future. In 2019, they hope to put kākāpō back on Resolution Island. It's a bold move because when conservationist Richard Henry did the same thing in the late 1800s, the stoats saw them off. The feisty mustelids are still there now, so it's not the perfect choice, but as the kākāpō population grows, it's a risk they're prepared to take. 'It's going to be an interesting and nervous experiment,' says Andrew. Kākāpō can be feisty birds so Andrew thinks they will cope, and if they do survive, it will be a fitting tribute to

Richard Henry. It was his dream to see kākāpō thriving on Resolution Island.

As we career towards the sixth mass extinction, it warms my heart to know that on a remote island, somewhere on the other side of the world, nest-minders camping out in flimsy tents are watching over the next generation of kākāpō. The work of the Kākāpō Recovery Programme demonstrates how, even when the outlook is bleak, we should never accept extinction as a fait accompli. When the Programme began, 25 years ago, the team had no way of knowing if they would succeed, but they had hope, determination and a bottomless bucket of ingenuity. They embraced science and technology, and ideas that were as eccentric as the kākāpō itself. Now it's paying off. The birds are still wild but we manage their world. We have become their nest-minders, their care-givers, their caterers and their match-makers. We have become directors of their evolutionary destiny, and as a result this avian oddity is coming back from the brink.

CHAPTER ELEVEN

Pigs and Purple Emperors

We heard them before we saw them. A rich, throaty rumbling reverberated through the undergrowth. From its volume, I could tell that the sound came from a big animal, but I didn't feel threatened. This was a happy noise. It made us all smile and quicken our pace to find the source of the gravelly grunting.

It was a warm summer's day as we picked our way through overgrown thickets of sallow and blackthorn. Straggly saplings stretched skywards amid a messy riot of dog rose and hawthorn. Dodging the whiplash of wilful brambles, we struggled to keep up with our tour guide, who had disappeared into the bushes. 'This way! Over here!' called the disembodied voice.

From time to time, the thickets gave way to sun-doused clearings where we'd pause and regroup. We stood knee deep in wild flowers. Butterflies danced across the fragrant blooms and the air was alive with birdsong. Without warning, a fallow deer startled and bolted from the undergrowth. Its distinctive black and cream rump was visible for a split second, and then the animal was gone.

I'm not sure how long we had been walking for. It was all too easy to lose a sense of both time and direction in this messy and magical wilderness, but our guide was sure-footed. Charlie Burrell seems to know every glade, pasture, tree, ditch and thicket of this varied and exuberant landscape. He can point out the trees where turtle doves nest when they arrive from Africa in the early summer. He knows the footpaths trodden by grazing herbivores and can identify the specific leaves that rare British butterflies use to lay their eggs. Along with his wife, the author Isabella Tree, Charlie has been responsible for the creation of this exotic wilderness, and now he rightly takes great pride in showing it off.

Along with a handful of other media folk, I'm visiting the Knepp Estate, a 14 square kilometre (5.4 square mile) jewel in Sussex, southern England. Part farm, part wildlife reserve, this is a place that exemplifies how people and nature can co-exist peacefully and sustainably. Knepp is a place where wild and domesticated species live side by side to their mutual benefit. It's a place that blurs the boundaries between wild and tame, farmland and wild spaces. It shows that there are alternatives to the industrial rearing of farm animals and that when we make space for wildlife – in our heads, in our hearts and on our land – everyone can reap the rewards.

In the previous couple of chapters, we've seen how conservationists guide evolution when they choose to focus on a particular species or type of organism. In Hawaii and elsewhere, scientists are selectively breeding coral in order to make them more resilient. In New Zealand, conservationists intensively manage the kākāpō because at present, they have no other option. In this chapter, however, I'd like to explore a very different type of conservation known as rewilding, where humans take a back seat and leave nature to get on with it. There is no focus on one particular species, and little in the way of predetermined goals. Instead, advocates of rewilding believe that if we give nature the space to grow, it can flourish without us. Rewilding is controversial and comparatively new, but the story of the Knepp Estate highlights its promise. Charlie and Isabella's actions teach us how sometimes sitting back and daring to be different is all it takes to help us engineer a brighter, more bountiful and biodiverse future.

We first noticed the noise after the feisty thicket had given way to a dusty bridle path. The source was close but obscured by the same obstinate flora we had just battled our way through. Ever the gentleman, Charlie held back a spiky bough and as we made our way past him, we could see a large, shady ditch surrounded by nettles and dead wood. The noisy beasts that had attracted our attention were kicking back in the muddy trench: a pair of blissfully bloated Tamworth pigs. Their russet tones blended beautifully with the earthy understorey. One of them lay snoozing with her head on a rotting log, while the other pottered around amicably. As the old saying goes, there's 'nowt as happy as a pig in muck'.

'I thought we'd find them somewhere round here,' said Charlie. 'They like to sleep in the middle of the day.' The pigs grunted happily, unfazed by their audience. 'These two are old girls,' he continued. 'They've been here long enough not to be bothered by any of us.'

Most other pig keepers would have slaughtered these elderly, barren sows long ago, but at Knepp things are different. Here the pigs are valued, not for their bacon, but for the ecological services they provide. Age does not stop pigs from rootling through the earth or creating a muddy mess, and it's this very talent that the Tamworths are prized for. Along with various other grazers, the Tamworth pigs have helped to make Knepp a biodiversity hotspot and a lifeline for endangered species.

Charlie bought the pigs in 2004, as part of an experiment he hoped would help save his then failing farm. He inherited the estate from his grandmother back in 1987 and in line with family tradition, decided to use the land for farming. He bought Holsteins and Friesians, and shiny new farm machinery. He grew wheat and barley, and doused the ground with fertilisers, fungicides and artificial hormones. He intensified and diversified, and launched the UK's answer to Häagen-Dazs: Charlie Burrell's Castle Dairy Luxury Ice Cream. Occasionally, he turned a profit. But as the global market expanded, the price of milk and cereal crops fell, and Charlie found himself battling against the farm's ultimate nemesis: its heavy clay soil. It's with good reason that the old Sussex dialect contains over 30 different words for mud. 'Gubber', 'ike', 'pug' and 'stoach' are all variants on the same claggy theme; an unavoidable assailant that stymied the farm machinery and made it hard for Charlie and Isabella to compete with farms that had better soils.

By the turn of the millennium, Knepp was haemorrhaging money and the farm was in crisis. 'We were losing cash for a long time,' says Charlie. 'We were really worried about the future, so we knew we had to change.'

Round about the same time, Charlie came across the work of a Dutch ecologist called Frans Vera. Vera is best known for his work at the Oostvaardersplassen, a Dutch nature reserve that is half an hour's drive from Amsterdam. Originally earmarked for industrial development when the site was reclaimed from the Zuiderzee over 50 years ago, it became a nature reserve by accident after economic recession hit and wildlife sneaked in. Marshy plants sprang up around the region's large shallow lake, attracting a rich and varied assortment of wetland birds. Many of them were rare species. In 1989 the Oostvaardersplassen was designated a Ramsar site: a wetland of international importance for nature.

The project showed how wildlife can flourish when people stand back and let nature take over, but Vera was concerned for the sanctuary's future. In accordance with the dogma of the day, Vera worried that unless there was some form of intervention, the reeds fringing the lake would creep towards the centre and smother the open water. Saplings would sprout and eventually the lake would be replaced by forest. 'What he was being told by scientists was that if you left nature to take its course, the site would revert to a closed-canopy woodland,' says Charlie, 'only that didn't happen.'

Instead, the Oostvaardersplassen became the 'go to' holiday destination for thousands of greylag geese, who flocked to the lake from all across Europe. They feasted on the reeds and marsh plants for the month or so it took them to moult, and prevented the lake from becoming overgrown. Thanks to the geese, sprawling reed beds were replaced with a more complex landscape of reeds and shallow water, which in turn attracted even more wildlife. 'Nature started to come back,' says Charlie.

So far so good, but there was another problem. How to persuade the geese to stay beyond the moulting season, so their benefits could be reaped longer term? Vera realised that if they could create the birds' usual habitat – grassland – next to the marsh, then the birds might stay, but rather than bringing in people to create the grassland, Vera had another idea. Human labour is expensive, but animals work for free. So he brought in sturdy Heck cattle, stocky Konik ponies and resilient red deer. Together, the herbivores manicured the region around the marsh, creating a lush, green pasture that was an instant draw for geese.

Now some 30,000 greylags – almost half the total population of north-west Europe – visit the Oostvaardersplassen every year. With their four-legged friends, the animals have created a patchy, mosaic habitat that is immensely popular with wildlife. Beavers, foxes, rabbits, water voles, stoats, weasels, polecats, grass snakes, beetles and butterflies have all found their way to the sanctuary. Wetland birds are thriving. Eurasian bitterns, spoonbills, great egrets and white-tailed eagles have all been spotted, and all in all, some 250 bird species have been recorded. The Oostvaardersplassen has blossomed to such a degree, it's now known by an alternative moniker: 'The Dutch Serengeti'.

Vera's experiences with the Oostvaardersplassen made Charlie realise two things: the value of keystone species and the value of letting go. Keystone species are species that have a disproportionately large effect on their surrounding ecosystem, relative to their numbers. Greylag geese, Heck cattle, Konik ponies and red deer are all keystone species because their actions create the conditions that enable numerous other species to thrive. When they are added to a landscape, they create and then maintain their characteristic ecosystem, and they boost biodiversity because the remodelled habitat attracts new life. They don't need much in the way of help. The story of the Oostvaardersplassen shows that if the right ingredients – or keystone species – are added to an environment, then nature can be left to sort itself out. Sometimes, letting go is all it takes. Charlie accepted that farming had become futile at Knepp, but now he found himself wondering if he could use keystone species to create his own nature reserve in Sussex.

At the start of the millennium, Charlie and Isabella auctioned off their farm machinery and began the slow process of relinquishing their farm to nature. They pulled down internal fences, re-seeded their fields with local grasses and wild flowers, and then began to introduce a judicious mix of large, free-roaming animals. With 'letting go' the new mantra, they chose hardy creatures that could look after themselves, and in time they welcomed fallow and red deer, old English longhorn cattle, Exmoor ponies and Tamworth pigs.

Tamworth pigs were chosen because this primitive breed is more closely related to its wild-boar ancestors than many of its modern counterparts. Watching the bloated sows kick back in the dirt, it's hard to imagine them as anything other than the rotund, slumbering slackers they seem to be, yet the Tamworth breed is surprisingly fleet on its trotters. Adult pigs can run as fast as a horse over short distances, so they can easily evade danger and look after themselves.

Exmoor ponies are another ancient breed. These are sturdy beasts with thick manes, strong legs and broad backs. In winter, they grow an additional insulating layer of wool underneath their water-resistant topcoat, and they have fatty pads around their eyelids that help to deflect the rain and snow. There's no doubting their credentials when it comes to self-sufficiency.

Similarly, old English longhorns are good at getting on with it. Just like their wild ancestor – the aurochs – they calve readily without the need for assistance, and can easily live outdoors all year round. Selective breeding has produced their thick brown and white coats, with a characteristic white stripe that runs down the spine, and a laid-back nature that seems at odds with their intimidatingly large horns.

Now each of these keystone species plays an important and distinctive role in Knepp's morphing landscape. 'They all have different behaviours and eat in different ways,' says Charlie. They have different mouths, digestive systems and food preferences, and eat different things at different times of year. Red deer, for example, like to graze in the spring and summer, but in later months when the plants they nibble become too tough or difficult to find, they start to de-bark trees. Cattle also graze, but when left to their own devices, they also browse twigs, brambles and saplings. 'They prefer to eat hedge than grass,' says Charlie. Through their actions, these herbivores variously destroy plants, move them around and create new habitats. Indeed it's been estimated that cows shift more than 200 different species around in their hair, hooves, guts and faeces. Pigs also eat plants and move them around but in contrast to the cattle, they adopt a less restrictive, more omnivorous diet. The Tamworth pigs are pot-bellied opportunists. When they're not snoozing, they spend their time rootling through the earth in search of invertebrates, rhizomes, roots and other tasty titbits. They eat the plants that other herbivores either can't reach or can't tolerate, like the roots of spear thistles and the poisonous fronds of bracken, and in autumn, they gorge on the acorns that rain down from the oak trees.

Almost 20 years after the first of these animals was introduced, Knepp has changed beyond all recognition. Where once there were fenced-in herds of dairy cattle and barren monocultures of maize, now there are free-roaming animals and a glorious, messy patchwork of different habitats. Hedgerows have spilled into fields and pasture blurs into scrubland, which blurs into woodland. The scrubland, in particular, has become a haven for wildlife. Far from the unproductive wasteland of common misconception, scrub – with its dog rose, bramble, hawthorn and sallow – offers browsing opportunities for the large grazing animals and new habitat for the wild plants and animals that have followed in their wake. Shielded by their tangled stems, saplings of oak, ash, wild service and birch use the thickets as nurseries. Now saplings can be seen emerging from their midst, creating more nurseries for invertebrates and sustenance for the creatures that feed on them.

The big mammals are crafting new opportunities for wildlife. Where they come to drink at the water's edge, they flatten the reeds and create space for other aquatic plants to grow. Where they stop to scratch on a low-hanging bough, their hooves compact the clay. When it rains, the depressions fill with water, providing new homes for invertebrates. Where pigs rootle, they create patches of exposed earth that attract burrowing bees.

Ants build their hills on these freshly ploughed clods, which in turn attracts insect-eating birds like wheatears and green woodpeckers. In summer, common lizards and small copper butterflies bask on the anthills, and grasshoppers lay their eggs on the surface. As the ants burrow, they change the composition of the soil, which has now become less acidic. As a result, new varieties of fungi, lichen, mosses, grasses and flowers are growing, but most astonishing of all is the effect that Tamworth pigs have had on one particular species: the Purple Emperor butterfly.

The Magic of Pigs

The Purple Emperor is one of Britain's rarest butterflies. The second-largest butterfly in Britain,* males sport an amethyst raiment decorated with ivory flecks and dark velvet eyespots, while females wear a mahogany-coloured variant on the same theme. Stunning, ephemeral and sadly in decline, these insects have long been on my hit list of 'must see' species, so when I discovered they were now thriving at Knepp, I was as happy as a Tamworth sow in a muddy ditch. Adult Emperors are on the wing between late June and mid-July, so when I planned my visit for 18 July, I was hopeful I would see one.

Early in the day, Charlie burst my bubble. 'I'm afraid the Purple Emperor season is over,' he said apologetically. 'We won't see them now.' To add insult to injury, he then mentioned that a week ago, a guest staying at the Estate's

---

*With a wingspan of around 10cm (4in), the largest British butterfly is the Swallowtail. The wingspan of the Purple Emperor is around 8 cm (3in).

glamping site had experienced a private display when a Purple Emperor visited him in the outdoor shower.

Male Emperors are gutsy creatures. One of the few British butterflies *not* to feed on nectar, the adults fortify themselves instead with fermenting tree sap and as a result, spend much of their time fighting in drunken duels. Rival males 'bundle' one another in territorial conquests that occur in the crowns of mature oak trees. Famously attracted to the smell of human sweat and stinky cheese,* there's no knowing exactly what attracted the male Purple Emperor to the shower that day, but Charlie was right. On the day I visited Knepp, there was not a single Purple Emperor to be seen. So instead, we visited their nursery.

Purple Emperors were once thought of as a woodland species, but Knepp is helping to rewrite the butterfly text books. When the species was first spotted on the Estate, back in 2009, they were found not in the closed-canopy oak forests, but in the scruffy scrubland I had battled to push my way through. As Charlie led us through the tangles and thickets, he paused in a clearing surrounded by sallow. Sallow is a type of willow tree that is often found in woodland, scrub and hedgerows. In Knepp, two varieties grow: common sallow (also known as grey willow) and great sallow (also known as

---

* Hardcore Purple Emperor fans – of which there are a surprising number – all have their own favourite lure with which to bait this ephemeral insect. Dirty nappies, fox poo, fish paste, roadkill, dog poo and rotting fish have all been tried, with more or less success. It's thought the butterflies are attracted to the salts and minerals they contain.

goat willow).* Sometimes, the two varieties hybridise, producing intermediate saplings with a variety of different leaf types. Charlie plucked a leaf from a nearby tree and turned it over in his hands. It was about the length of his thumb, and tapered at both ends. 'Purple Emperors lay their eggs on these young hybrid saplings,' he explained. 'The saplings produce lots of different leaves, but it has to be this one particular form – not too hairy, not too waxy, not too small, not too big, not too long and not too thin.' He placed the leaf in the palm of his hand and held it out to show us. 'This one is just right,' he exclaimed.

Like so many infants, Purple Emperor caterpillars are fussy eaters. Only a small proportion of the leaves produced by these hybrid sallow saplings are deemed good enough for the offspring of the gravid females. Since shrubland and sallow have been allowed to thrive at Knepp, sightings of the Purple Emperor have soared. On a single day in June 2017, 148 individuals were spotted. It's a staggering amount for a species that is scarcely seen, and Knepp is now the species' number one UK breeding hotspot. But where do the pigs fit in?

Charlie explained that the hybrid saplings have to be young, so a constant supply of freshly germinating plants is needed. 'The seed rains down from the sallow for a couple of weeks in May but they need to fall on open earth in order to germinate,' he says. 'Pigs are like ploughs. They rootle around and open up the ground.' Without pigs, or some other form of disturbance, there would be

* Both varieties are known by the more familiar name, pussy willow.

no freshly emerging hybrid sallow trees, and without them, there would be no Purple Emperors.

'That's the magic of it all,' Charlie enthused. 'When we started all this, we never realised there would be a connection between the Purple Emperor and the pig, but there it is!' The pigs created the opportunity for the Purple Emperor to thrive. Similarly, it's thought that the Tamworths' rootling is creating the perfect conditions for certain other weeds that are the preferred foodstuff of turtle doves during the summer. This could be at least part of the reason why turtle doves, once teetering on the brink of extinction, are now doing so well at Knepp. The Estate now has Britain's fastest growing population of turtle doves. They might not look like your conventional conservation heroes, but Knepp's Tamworth pigs have arguably done more to aid the rescue of endangered species in Sussex than any human being.

Letting Nature Run its Course

The Knepp Estate is a beautiful example of the power of rewilding. Rewilding is all about creating more wild spaces so that wildlife can flourish, and people can reconnect with nature. Alastair Driver, Specialist Advisor to the UK charity Rewilding Britain, describes rewilding as 'the large-scale restoration of ecosystems to the point where nature is allowed to take care of itself'. In its early stages, rewilding needs human help. If the aim is to create woodland, for example, then trees need to be planted. If the aim is to create paths for migratory fish, then dams need to be razed. If farmland is being rewilded, then its soils need to be purged of pesticides. Over time, as natural processes kick in, human involvement can then be reduced and nature can be left to look after itself. 'Rewilding is an ongoing activity,' says Alastair. 'It takes intervention to start with, leading to a long-term reduction in activity and management. This could take decades or centuries or more.' It's a long-term commitment but it's worth it. Rewilding has the potential to increase both the range and number of many flagging species and as such, can be used to influence the course of evolution. It gives us the opportunity to undo some of the environmental damage we have done, and steer our planet towards a greener, more biodiverse future. Now rewilding projects are popping up all over the globe.

Reintroducing animals is a key part of the process. If a species has become extinct locally, then representatives from another place can be recruited. In 2009, for example, beavers were reintroduced to Scotland from Norway after a 400-year absence.* Now the animals are beavering away remodelling the woodland on the banks of the region's lochs. If an animal is extinct globally, then experts may choose to substitute a closely related proxy. The Snares Island snipe, for example, was used as a stand-in for its extinct relative, the South Island snipe, when the little sandpiper-like bird was introduced to New Zealand's Putauhinu Island. At Knepp, Charlie and Isabella substituted living species for extinct ones – English longhorn cattle replaced the extinct aurochs and Exmoor ponies stood in for tarpans – but sometimes if

* They were hunted for their dense pelts and for castoreum, a yellowish secretion from scent sacs near their tail that was once used to make perfume and medicine.

there is no closely related proxy, rewilders may consider an alternative based on ecology. Long before Polynesians arrived, the Hawaiian island of Kauai was home to a giant, flightless duck called the moa-nalo. Recently, when conservationists were trying to restore part of this island ecosystem, they realised that the duck's extinction had left an ecological hole. A large, gentle grazer was needed, so they introduced giant tortoises, which help to control the spread of invasive plants and promote the growth of native flora.

In continental Europe, the idea of rewilding has been widely embraced. A group called Rewilding Europe seeks to rewild 10,000 square kilometres (3,900 square miles) of land, an area the size of Cyprus, by 2020, and is encouraging other organisations to rewild a further 100,000 square kilometres (39,000 square miles), an area the size of Hungary. From the wooded savannas of Western Iberia to the reed beds of the Danube Delta, from the forests of Lapland to the rocky slopes of the Rhodope Mountains, Europe is becoming wilder. Large, charismatic species are returning to these and other regions. Tauros cattle have been released into the forested valleys of Croatia's Velebit mountains (see Chapter 2). European bison (or wisent) – impressive beasts that once ranged from central Russia to Spain – have been returned to Germany, Spain, the Netherlands, Denmark and parts of eastern Europe. Beavers have been released on more than 150 occasions, and golden jackals, wolf-like canids once purged from Europe, are now breeding in Bulgaria, Hungary and the Balkans. Bears now live in more than 20 European countries and wolves have spread across most of the continent.

The impact on wildlife has been startling. Little by little, centuries of ecological damage are being unpicked, with the benefits to people being equally wide-ranging. We all know that trees absorb carbon dioxide and provide us with oxygen, but they also absorb huge amounts of rainfall. One tree-planting scheme in Wales has revealed that, where there are trees, water drains into the soil at 67 times the rate that it does when there is just grass. Forests help to prevent potential flooding. So do beavers. Despite the concerns of British farmers, the general consensus is that the canals and dams built by beavers act as sponges, soaking up excess water and reducing flood risk. Their dams also trap silt and help to purify the freshwater that often ends up in our reservoirs.

Who doesn't want to see a beaver dam or catch a glimpse of the elusive Eurasian lynx? These vastly improved wild spaces are a draw for tourists, which in turn generates jobs and income. National parks and nature reserves around the world host around 8 billion visits per year, contributing around £450 billion ($565 billion) to local economies. And it soothes the soul. Being in touch with nature helps people to connect with the planet. There's a sense of wellbeing that cannot be gleaned from bricks and mortar. A study from the UK's mental health charity, Mind, confirms what most of us know intuitively: that interacting with the natural world is good for mood and self-esteem.

Of course, the idea of setting land aside for nature is nothing new. People have been protecting species and their habitats for well over a century. Reintroductions have been happening for a long time too, but rewilding is different from traditional conservation. In the past, conservationists focused on collecting wild spaces like postage stamps. We have nature reserves and protected pockets of land, but these tend to be managed and fenced in. There's no denying they're important, but on their own, they are tiny oases in a desert of sprawling urbanisation. In contrast, rewilders imagine a different scenario involving vast networks of land set aside for nature, and it is here that the controversy begins.

Critics of rewilding caution that the reintroduction of lynx, bison, beaver and other animals could have unforeseen consequences. They suggest that if the reintroduced species do too well, they could become problematic. Instead of nurturing native wildlife, they could displace it. As farmland becomes repurposed for rewilding, farmers worry that their livelihood could become eroded. They caution that their livestock could be at risk from introduced predators like wolves or lynx, and that nearby beaver dams could cause flooding that would damage their crops. In 2013, rewilders faced a barrage of criticism when the large grazing herbivores of the Oostvaardersplassen were left to fend for themselves during a particularly harsh winter. In line with the principles of hardcore rewilding, the rangers at the Dutch nature reserve didn't give the animals any extra food and as result, thousands of animals starved. It led to protests and fraught scenes as people hurled hay bales over the fences into the reserve. Ecologists and rangers received death threats, and some critics even compared the nature reserve to Auschwitz. In the midst of the controversy, more than half of the reserve's red deer, Konik horses and Heck cattle had to be euthanised by the Dutch state forestry organisation.

The Oostvaardersplassen has been divisive. Some conservationists see the thin and starving animals as part of the natural cycle of life, while others find this hard-line approach disquieting. Now, as we humans increasingly manage the natural world, we need to decide what level of intervention we are comfortable with. Should the natural world – famously red in tooth and claw – sometimes be left to get on with it or have we become long-term curators of the planet's wild spaces?

Meanwhile, rewilding continues to court controversy. When people talk about returning beavers, deer or cattle to the landscape, it's one thing, but when they talk about introducing much bigger animals, the stakes appear to be raised. Ladies and gentlemen, it's time to talk about the elephants in the room.

Emancipation for Elephants

'When we first told people about our idea, the initial reaction was varied,' says ecologist Jens-Christian Svenning from Aarhus University, Denmark. 'Lots of people thought it was pretty crazy, to be honest.' 'Some people laughed at us,' says his colleague Ole Sommer Bach, who is the curator of Randers Rainforest Zoo, also in Denmark. 'They thought it must be a practical joke or some kind of provocation.' But the two men were deadly serious. Jen and Ole set tongues wagging when they suggested that elephants be reintroduced to Denmark.

In 2013, the duo wrote a paper outlining their proposal. They imagined introducing a small group of Asian elephants to northern Denmark, where the animals would be allowed to live in large, fenced enclosures. 'But we wouldn't just set them free,' assures Ole. 'This would be carefully controlled.' Ole knows what he is talking about. He has form when it comes to rewilding. In 2010, he oversaw the successful reintroduction of bison to Denmark after an absence of around 8,000 years. Now he thinks elephants could literally be the next big thing.

'The enclosure would be as big as possible,' he says, 'ideally with a mixture of habitats including meadow and forest.' The animals and their actions would be monitored closely, and then if the public agree, in time the number and range of the elephants could be increased. It would be different from keeping them in a zoo or safari park, because the elephants would have the space and scope for a wild life. They'd be able to choose where they go, what they eat and how they spend their time, without the restrictions of a more confined environment. Crucially, humans would step in from time to time if assistance was needed. The famine of the Oostvaardersplassen casts a long shadow, so the elephants would never be allowed to starve. 'Some people say there should be no human involvement at all,' says Ole, 'but I'm a sissy. I come from a zoo. I cannot see the point of animals starving to death.'

If your knee-jerk response is that wild elephants simply don't belong in Denmark, then think again. We're conditioned to think about the short-term past – events that have occurred in our lifetime or further back in recent history – but when you look back across geological time, elephants were a common, almost constant feature of the continent. 'Europe has had elephants on the mainland continuously for the last 18 million years, right up to 10,000 years ago. They are long-term members of temperate ecosystems in Europe,' says Jens. Modern humans, in contrast, arrived in Europe around 40,000 years ago. So if anyone doesn't belong here, it's us!

During this lengthy European residency, there was no 'European elephant'; rather there were waves of elephants that colonised the continent from Asia. 'For large animals, Europe is just a peninsula on Asia,' says Jens. Elephants from the east outcompeted and replaced elephants in the west. It was a dynamic process unconstrained by modern geographical boundaries, so if the idea of importing Asian elephants to Denmark floors you, remember this. Asian elephants have been introducing themselves to Europe for many millennia.

From a natural-history perspective, there are multiple reasons to believe that the reintroduction of elephants could have positive effects. Just like greylag geese and Tamworth pigs, elephants are keystone species that have a major impact on their environment. Elephants are grazers, browsers, seed dispersers and bulldozers. They rip up and knock down trees, help constrain the spread of woodland, and maintain and fertilise large stretches of grassland. When they disappeared from Europe at the end of the Pleistocene, the services they provided also disappeared. Now there's a big elephant-shaped hole in the landscape. If they were returned to Denmark, and these processes were restored, elephants would create opportunities for other species to thrive.

In Denmark, there are huge areas of grassland which could be used to sustain them. At present, these areas are maintained by people and by traditional grazing animals, but it's expensive and time-consuming. 'When we maintain nature in municipality in Denmark today, it's seen as a cost,' says Ole. 'It's seen as a problem. We think it could be done easier, more cheaply and more spectacularly by the use of elephants than by commercial farm breeds or by machines. Instead of problems, we create possibilities. There could be a whole tourist economy connected to it.'

'At the same time, there's a global perspective,' Ole continues. Poaching, conflict and habitat destruction are pushing the world's largest land mammal to the brink. The number of Asian elephants has been reduced by half over just three generations. There are now less than 50,000 left. So the question is, should we leave it to Africa and Asia to save the elephant? 'I don't think we can,' says Ole. 'If they go extinct, it's a global problem.' Increasingly, elephants are not safe in Asia or in Africa, so perhaps it's time to start establishing wild populations elsewhere. That way, if elephants go extinct from their current range, there will still be a wild backup.

It's a scenario borne out by the banteng, a species of wild cattle that hails from Southeast Asia. In 1849, the British military brought banteng to the Cobourg Peninsula in northern Australia where they used them for meat, but when the outpost was abandoned, the banteng wandered off into the wild. Since then, they have slotted into the native ecosystem and the Garig Gunak Barlu National Park where they now live is none the worse for their presence. The banteng have even struck up a relationship with the native Torresian crow, which picks ticks from their fur. It's the first documented case of a native species forming a mutually symbiotic relationship with a non-native one.

Meanwhile in Southeast Asia, the banteng is now endangered and conservationists are considering whether or not Australian banteng could be used to help boost their numbers. One hundred and fifty years ago, no one could have predicted how important the Australian banteng population would become. Now the expat animals have become part of an accidental rewilding success story that demonstrates the value of setting up insurance populations of endangered species away from their conflicted native homes.

There could be other reasons to relocate elephants too. Ecologist David Bowman from the University of Tasmania has suggested that elephants could be used to help control invasive species. In a 2012 *Nature* opinion piece, he suggested introducing African elephants to Australia, of all places. I say 'of all places' because Australia is a continent famously overrun by invasive species: opportunists that seized ecological niches left vacant by the demise of the country's own megafauna some 50,000 years ago. It's a bold character that suggests introducing one non-native species to deal with another, but then gamba grass is a big problem.

Gamba grass is a giant, tufty African grass that was originally introduced to Australia in the 1930s as a pasture grass. In 1986, the Northern Territory Department of Primary Industries developed a cultivar that was, in their own words, 'easily established', 'highly productive', 'drought resistant' and 'adapted to a wide range of soils'. They put it in some trial sites and promised to contain it, but the grass escaped. The characteristics that made it a good pasture grass also made it an aggressive weed. Now it is one of the world's most rapidly spreading invasive plants and currently affects up to 15,000 square kilometres (5,800 square miles) of Australia's Northern Territory, where it is outcompeting the native savanna grasses. This is a problem. Gamba grass grows up to four metres tall and produces up to five times the biomass of native Australian grasses. In a continent increasingly ravaged by wild fires, this has proved calamitous. Australia's savanna trees have evolved to cope with low-intensity fires that are close to the ground, but they struggle to survive the intense infernos caused by burning gamba grass. It's not only the trees that burn. The threatened native marsupials that shelter in them burn too. Wildlife takes a hit and over time woodland becomes replaced by pasture. This is an ecosystem wildly out of sorts.

Bowman thinks that elephants could help set things straight by consuming the gamba grass and slowing its spread. Australia's large grazing animals – kangaroos, cattle and buffalo – aren't up to the job, and alternative solutions involving herbicides and manual labour are expensive and damage the delicate savanna ecosystem. Using mega-herbivores to control gamba grass may seem radical, says Bowman, but it could be more practical and cost effective than other strategies.

Big public ideas require big public support, so for now these pachydermal propositions remain at the planning stage. No one is about to release elephants into Australia or Denmark any time soon, but the ideas are at least worth considering. Elephants *could* help boost biodiversity and curb invasive grasses. Their presence *could* generate income from ecotourism, and provide a safe haven for a species that is not safe in its current natural range. And if relocating elephants from Asia to

Europe feels like too much of a stretch, Ole has an alternative.

'In Denmark, we still have elephants in circuses,' he says. Wouldn't it be wonderful if we could set them free? 'It's a question of, do we want to keep these animals in small enclosures and look at them and throw popcorn at them, or do we want to use them in a larger landscape that's screaming to have mega-herbivores back? I'm sure that if the elephants could make a choice, I know what they would say. It would be nice to use the last circus elephants for the first rewilding project.'

Maybe introducing Asian elephants to Denmark or African elephants to Australia isn't such a crazy idea after all, but what about woolly mammoths? In Siberia, a father-and-son duo called Sergey and Nikita Zimov believe that rewilding the frozen north with megafauna could help the world to combat one of its biggest problems: climate change.

A Shaggy Elephant Story

In the woolly mammoth's heyday, Siberia was a very different place. The large, shaggy beasts grazed in lush, open pastures that covered much of Eurasia and North America. Vast herds of mammoth, bison, reindeer and horses roamed the plains, under the watchful eyes of cave lions and wolves. The so-called mammoth steppe was a biodiverse and vibrant ecosystem. 'It was like an Arctic Serengeti,' says Nikita. 'Modern humans didn't need to worry about finding food; they needed to worry about being trampled.'

Then, when the Pleistocene came to an end 11,700 years ago, it all disappeared. The mammoth steppe was replaced by scraggy, unproductive tundra. Much of the megafauna, including woolly mammoths and aurochs, went extinct and the grasslands that they lived on became entombed in permafrost. Scientists now estimate that 1,400 billion tonnes of organic carbon lies hidden in the permafrost, the frozen sub-surface layer of soil, ice and rock that covers around a quarter of northern hemisphere land. That's roughly twice as much as exists in our atmosphere and three times the amount found in all the world's forests combined. Now, as our world warms and the permafrost melts, microbes are starting to convert this organic carbon into methane and carbon dioxide. The concern is that as these greenhouse gases bleed into the atmosphere, they will accelerate the rate of global warming, leading to more melting and microbial activity. It's been likened to a ticking carbon time bomb. 'We are rapidly approaching the point where the Arctic permafrost will start thawing everywhere,' says Nikita. 'This could be catastrophic.'

The Zimovs believe that if large grazing animals are returned to the far north, they could help to slow this warming. Sergey first outlined his ideas in a Soviet journal in 1988. He explained how the mammoth steppe ecosystem was shaped by the large animals that lived there during the Pleistocene. In the summer, large herbivores maintained the grasslands by eating weeds, dispersing seeds and fertilising the ground. By knocking down trees and eating their bark, they also helped to keep forests at bay. Grassland is lighter than forest and reflects more sunlight back to space, so the pastures they maintained had a cooling effect. Similarly, in winter, when the animals trampled the snow to find fodder, they exposed the earth to the bitter Arctic air. Put simply, big grazing herbivores helped keep the Arctic cold. If they could be returned, he mused, they should be able to convert the mossy tundra back to productive grassland *and* help to keep the permafrost frozen.

In 1996, Sergey set up an experimental nature reserve to test his hypothesis. The reserve, which occupies around 20 square kilometres (8 square miles), lies close to the Kolyma River in the Sakha Republic in north-east Siberia, and in a nod to Jurassic Park, he called it Pleistocene Park. With no external funding, in the early days it ran on fumes and enthusiasm. First, he acquired some stocky, semi-domesticated horses from local Siberian natives who kept them for meat; but without any fences, the animals just wandered off. So then he added fences.

Today Pleistocene Park is home to around 30 sheep, 30 reindeer, 9 yak, a few musk ox, one bison and a couple of dozen horses, all species that used to live in Siberia during Pleistocene times. Now shrubs once so tall that they overshadowed people have been grazed to waist height or less. Tussock, a common, slow-growing weed, is giving way to meadow grass. Slowly but surely, the mammoth steppe ecosystem is starting to return. 'It's the start of a long process,' says Nikita, 'but there are encouraging signs.'

Crucially, experiments performed at Pleistocene Park have shown that where big herbivores graze, soil temperatures are, on average, several degrees cooler than where grazers are absent. This is tantalising evidence that the Zimovs' approach really could keep the permafrost frozen. And the Zimovs believe that the carbon-sequestering skills of the animals they introduce will hugely offset the carbon footprints of the individual animals.

If the woolly mammoth can be de-extincted (see Chapter 4) then the Zimovs would welcome them to Pleistocene Park. What sets mammoths aside from other large Pleistocene grazers is their ability to uproot, knock down and generally destroy woodland. Because forests absorb more warming solar radiation than grasslands, keeping the trees at bay is one of the Zimovs' key goals. Mammoths are the perfect creatures to do this.

In the meantime, the permafrost is slowly, patchily starting to thaw and in Siberia the consequences are all too apparent. Enormous sink holes are appearing and huge chunks of land are falling away. Most buildings in the region rest on permafrost, but with their foundations slowly melting, entire houses are sinking into the mud. Nikita thinks that Pleistocene Park's nearest town, Cherskii, will collapse in the next 30 years, displacing all of its 2,500 residents.

'We're not losing a lot of carbon from the permafrost right now,' says permafrost scientist Max Holmes from the Woods Hole Research Centre, Massachusetts, 'but we are headed in that direction.' If we continue to burn fossil fuels at the rate we do now, we stand to lose around 1.5 billion tonnes of carbon per year for the rest of the century. 'That's like adding another USA into the mix,' says Max. Pleistocene Park, and its extended network of Ice Age nature reserves, could help slow the loss. 'I think Pleistocene Park could make a difference,' says Holmes. 'I don't think it's *the* solution to climate change, but I think it could be a small part of the solution.'

The Zimovs, however, aren't putting things on hold, waiting for the herds of de-extincted mammoths that would be needed to help manage the region's forests. They believe they can recreate the mammoth steppe ecosystem without mammoths.

In the past, both Sergey and Nikita mimicked the actions of mammoths by mowing down trees in a battered old tank that they imported from the Ukraine. It took them two months to drive the khaki-coloured boneshaker all the way from the Russian border to the park. 'We used it to knock down trees when we were building fences, and when we were entertaining journalists,' says Nikita, 'but it's broken now and we don't want to change the ecosystem with a tank.' Tanks aren't green, efficient or sustainable, so Nikita would rather have the animals do the job.

So for now they are pinning their hopes on bison. Like woolly mammoths, bison are a keystone species. They can fashion entire ecosystems by trampling, grazing and fertilising. And they kill trees, not by knocking them down, but by eating their bark. Until recently, Pleistocene Park had only one bison, a male that came from Europe, but now he has been joined by a mini herd that made the long journey from Denmark. The Zimovs imagine a huge interconnected network of nature reserves bursting with large, grazing animals, including bison, horses and cattle. 'We need to take huge territories,' says Nikita. 'This is a century-long plan.' Immense numbers of animals will need to be imported, including predators to keep the grazers in check. 'We have bears, wolverines and polar foxes but we need wolves and maybe Siberian tigers,' he says with a twinkle in his eye.

The Zimovs are a force to be reckoned with. Their vision is big and bold, but in a world beleaguered by catastrophic climate change, where is the harm in letting them try? The Arctic is remote and scarcely populated. It's the perfect place for a rewilding experiment of this nature, and any valuable lessons can then be applied to other rewilding projects elsewhere.

Wilding the Tame

In time, as these projects unfold, we will learn the value of letting elephants live wild in Denmark, and of stocking the Arctic with herds of big, grazing herbivores. Meanwhile, ongoing rewilding experiments, like those at Knepp and the Oostvaardersplassen, hint that when humans step in and give nature a helping hand, wildlife can blossom. This laid-back conservation approach brings big gains. Twenty years after Charlie Burrell and Isabella Tree sold their farm equipment and embarked on their rewilding foray, Knepp has become a Mecca for wildlife. Thanks to their relaxed, permissive attitude and the keystone species they introduced, the site now boasts an impressive list of native and immigrant species. Critically endangered turtle doves and nightingales now breed here, as do peregrine falcons, sparrowhawks, lesser-spotted woodpeckers, lapwings, skylarks, woodlarks, yellowhammers and woodcock. All five UK species of owl are found here, as are 13 of the UK's 17 breeding bat species. Nightjars – more usually associated with heathland – have put in an appearance, as has the occasional black stork, one of the rarest birds in Western Europe. There are foxes, polecats, stoats, badgers, hedgehogs and mink. With no pesticides or fungicides to thwart them, fungi, mosses and beetles are thriving. 'I have found 27 species of dung beetle in one cow pat,' Charlie told me with pride. 'What we are finding is huge numbers of everything. There are lots of scarce and wonderful things.' This is all the more remarkable when you consider that Knepp is less than 50 miles from London, and less than 20 miles from the UK's second busiest airport, Gatwick.

Charlie and Isabella have shown that when large herbivores are left to their own devices they do amazing things. At Knepp, they are helping to reverse the decline in British wildlife. Along the way, Charlie and Isabella have managed to create a sustainable business. They now offer wildlife-spotting safari tours and butterfly-laden glamping, and when the herd gets too big, they sell their surplus Longhorn cattle for high-quality, ethical, organic beef. Knepp is proof that farming doesn't have to be intensive, fuelled by pesticides and antibiotics, and occur at the expense of local wildlife. Simply by introducing a few carefully chosen keystone species, it's possible to promote the creation of rich, beautiful landscapes that can support the production of high-quality, ethically sound cattle and wildlife alike. For tens of thousands of years, we have been taming the wild. Now Knepp and other projects like it are beginning to show us the value of wilding the tame.

CHAPTER TWELVE

The New Ark

For most of the last 3 billion years or so, life on Earth was shaped by non-human forces. Evolution tended to happen slowly, with species developing across millennia. Then along came a bolshie, bipedal primate that decided to call itself 'human'. The Earth's natural history came to an end. In its place arose a post-natural era where the fate of all living things became irrevocably intertwined with our own.

Around 750,000 years ago, our ancestors began systematically butchering large animals. Around 500,000 years ago, they worked out how to attach sharpened stone points to sticks, in a process known as hafting. Newly armed with spears, their hunting prowess increased significantly and their prey fell more readily. Around 300,000 years ago, our species, *Homo sapiens*, emerged in Africa, and then began to spread around the globe. Fifty thousand years ago, we started playing with fire. Studies of charcoal and pollen records suggest that when early humans arrived in Borneo and New Guinea, they used flames to help clear the land. All these interventions altered the environment, but none had the impact of what would come next.

Forty thousand years ago, when we made the wolf roll over, our relationship with the environment moved into a new phase as we began to deliberately alter the species we lived alongside. The domestication of the wolf is a pivotal moment because this was the first time that humans intentionally changed one sort of animal into another. After the wolf, other species followed. We domesticated crops and fashioned farm animals. A few hundred years ago, farmers started selectively breeding their animals, resulting in the specialised breeds that we have today. Seventy years ago, breeders started using artificial insemination as a way to spread desirable genes through populations of farm animals, and today the process has become further nuanced by the advent of genetic testing and cloning. Then, as molecular tools came on board, we began to sculpt genomes with increasing levels of sophistication. The genetic makeup of living things can now be altered with pinpoint precision, enabling us to further refine the organisms we create.

Domestication was the bedrock that enabled civilisation to flourish. Domestic species fuelled the growth of cities, trade and the settled lives we now enjoy. Domesticated animals and plants nourish the world. They work for us, they live alongside us and they give us products like leather, wool and cotton. We have created long-haired guinea pigs, giant rabbits and dogs so small they can fit inside a handbag. We've made goldfish with fancy fins and domestic birds in every colour of the rainbow. We've cloned polo ponies. We have toyed with hybrids, deliberately interbreeding species in order to generate novelty. Animals have been repurposed as bioreactors. Scientists have tweaked the DNA of mammals, fish and fruit flies to make animal models of disease, raising hopes for the development of new therapies for devastating disorders.

We currently find ourselves in the midst of a complicated relationship with the organisms we have created. Things have become desperately one-sided. We have a tendency to modify animals and plants for our benefit, rather than theirs, and turn a blind eye to the suffering that can sometimes ensue. Modern broiler chickens may be meaty, but they struggle to stand. Dairy cows may supply us with vast quantities of milk, but they frequently end up lame. Some domestic breeds are now so big they can't give birth naturally. Meanwhile, in our homes, we welcome pet breeds with congenital health problems that have been caused by selective breeding.

This is the antithesis of natural selection. According to natural selection, useful features and their genetic underpinnings become propagated if they make a species more likely to survive and reproduce. In contrast, useless features and their genetic underpinnings are not selected for, and so tend to disappear over time. In recent times, it has become commonplace to engineer species with characteristics so intemperate, they confer no advantage to the organism. The changes are at best neutral, at worst harmful. All too often, we prioritise our own needs, desires and whims over the welfare of the species supposedly in our care. This is not a kind or sustainable view.

In farming, the use of artificial insemination means that key animals now dominate the gene pool. As a result, some industrial breeds are becoming genetically homogeneous and inbred. This does not bode well for their long-term future. Meanwhile, the widespread adoption of highly productive breeds, such as Holstein cattle, has been at the expense of more traditional breeds that have fallen from favour. These primitive farm breeds contain unique genetic variants that could prove useful in the future, but now many are endangered and many more have already gone extinct. We lose this valuable resource at our peril.

There are now 70 billion farm animals and the vast majority are industrially reared in factory farms. Instead of letting them graze, browse and forage, we give over vast swathes of land in order to grow crops to feed them. It's perverse. Wildlife on one side of the globe is being destroyed so we can grow crops that are fed to domestic animals on the other. Now a quarter of all mammals, an eighth of all birds and more than a third of all amphibians are endangered, and every day more than 30 species go extinct. Human activity is to blame and factory farming plays a big role. The way we manage our domestic species has become a key driver of the sixth mass extinction.

When my children were small, someone gave them a toy ark. We may well be atheists, but the ark was appreciated and it got played with nevertheless. It was the usual: a robust wooden boat with an ample deck and plenty of space in the hold. As you would expect, the animals came in twos. The set was supplied with matching pairs of elephants, giraffes, monkeys, camels, zebra, lions and crocodiles. Oh, and a couple of doves. They boarded the boat via a rickety gangplank, watched over by a pair of wooden biblical figures – voiced by my children – who warned the animals *not* to eat each other or 'things could get nasty'.

Consider the wildlife on Earth today. If someone were to make a new ark that correctly reflects the current abundance of animal life, it would be very different from my children's old-school play set. The balance of life on Earth has shifted radically since modern humans first came on the scene. Ten thousand years ago, 99.9 per cent of the world's land-living mammalian biomass was comprised of wild animals. Today, 96 per cent is made up of domestic animals and people (in a 2:1 split). Wildlife makes up the remainder. While livestock abounds, wild species barely get a look in. If the shipmates of the new ark were made as tiny wooden replicas and allowed to board in pairs, the roll call wouldn't go 'elephant, giraffe, monkey, camel, zebra, lion, crocodile'. It would go 'chicken, chicken, chicken, chicken, chicken, chicken, cow.' This distribution of life is not sustainable.

It's hard for us to grasp the magnitude of the transformation that has been caused by our changing relationship with the natural world. This is partly because much of the natural world is remote – out of sight, out of mind – and because many species disappear quietly, without fanfare or formal identification. It is also because we have a generational blindness to long-term change. This is called 'shifting baseline syndrome'. We look at the ecosystems that surround us today, and then compare them with the only baseline we know: the world that existed when we were young. I frequently lament the lack of butterflies in my garden and hark back to my formative years when our straggly buddleia was smothered in flying insects. This personal baseline, rooted in the past, is perceived as the default setting, but butterflies were already in decline when I was a child. So although we may notice when wildlife diminishes, we fail to realise that the baseline we judge this loss against is already a state of impoverishment.

Thanks to these shifting baselines, we struggle to comprehend what the world was like more than a single generation ago. We have no idea that the default setting for most ecosystems is one of abundance, chock-full of megafauna. When London's Trafalgar Square was excavated in the nineteenth century, builders found the remains of hippos, straight-tusked elephants, giant deer, aurochs and lions. Where there are pigeons and tourists today, in the past – a little over 100,000 years ago when the climate was similar to the present day – there were megabeasts. I'm not suggesting we return elephants to Trafalgar Square, but is it really so far-fetched to imagine that the pachyderms could live a wild life in Denmark's national parks, or that bison could live in Siberia? As they are keystone species, the ecological benefits would be immense.

Rewilding projects, such as the Knepp Estate and the Oostvaardersplassen, highlight the ecological value of adding large, grazing animals to the landscape, and show us that farming doesn't have to be intensive to be profitable. Great things happen when you let domestic animals be wild. At the Knepp Estate, the Tamworth pigs have been spotted diving for swan mussels in the silt at the bottom of the lake. When it's time to give birth, the English longhorn cattle sidle off into the long grass and get on with it. A little while later, the new mothers find nettles to nibble because, it's thought, it helps to replenish their depleted iron reserves. When they are left to be wild, domestic animals express wild behaviours. It's wonderful to see. These are unquestionably content animals that are in tune with their natural environment. They live their lives with minimal human interference, and as keystone species, create opportunities for other species to thrive. At a time when biodiversity is plummeting, it's a wonderfully laid-back and successful approach to conservation.

When conservation first emerged as a discipline, around 150 years ago, it marked a fundamental shift in human attitudes. It was the first time that humans began to systematically prioritise other species over our own. We began to think about the value of nature, and the needs of wildlife. In the late 1800s, when the New Zealand conservationist Richard Henry relocated his beloved kākāpō to their remote island sanctuary, he didn't do it for himself. He did it because he realised that without his assistance, the birds might not survive. It's that same altruistic attitude that motivates the members of the current Kākāpō Recovery Programme, and indeed all conservationists the world over.

We live in turbulent times and yet, perversely, we've never been better equipped to deal with the biological losses we are facing. Rewilding is just one of many options. If we can clone horses, perform coral IVF and decode the genomes of every single kākāpō, then just imagine what else will become possible in the near future. We stand on the shoulders of scientific giants, and as technology improves, I am sure that we'll find solutions to some of the ecological problems that are currently intractable. Scientists are using molecular methods to genetically modify domestic animals with great success, but why stop there? Instead of modifying animals for our benefit, isn't it time we started modifying them for theirs? Although it may seem like an outlandish idea, I believe there are circumstances where the deliberate genetic modification of wildlife is justified.

Consider the black-footed ferret, a North American mustelid that was driven to the brink of extinction by disease. Thanks to a successful captive breeding programme, thousands of ferrets have been reared and released into the wild, but all are descended from just seven founder members. They are desperately inbred and still vulnerable to the disease that originally blighted them, so they're not completely safe yet. Scientists have cells from two historical animals that died without reproducing, so they'd like to use these cells for cloning and then add the resulting ferrets to the current breeding pool. It would effectively boost the founding population from seven to nine, which could help to reverse the effects of inbreeding. In addition, they'd like to use CRISPR-Cas9 to edit their genomes so they become resistant to their nemesis, the sylvatic plague.

It would be like adding a sprinkling of well-timed, beneficial mutations to the current evolutionary churn. The peppered moth sailed through the Industrial Revolution because a newly acquired mutation helped it to adapt to its sooty surroundings. It was lucky: the random genetic blip appeared at just the right time. But we can't expect other species to be equally blessed. If we sit back and wait for the black-footed ferret to acquire a similarly fortuitous mutation, we could be waiting a long time. With CRISPR-Cas9, we could engineer a genetic solution overnight.

Other species could benefit too. The Kākāpō Recovery Programme demonstrates the value of technology and bold thinking. Perhaps one day, CRISPR-Cas9 could be used to engineer disease resistance or restore lost genetic diversity back into the emerging kākāpō population. If it helps the species to survive, then it could be worth considering.

There are caveats. This is not a technology to be wielded lightly. This is emergency triage for conservation. If only we had cared more for the natural world in the first place, then this discussion would be superfluous. Yet here we are. Gene editing won't be *the* solution to the dwindling levels of biodiversity, but it could be *part* of the solution. Conservation needs more tools in its armoury, so this is at least worth exploring. Rewilding, assisted evolution and intensive management strategies are all part of the picture. It may seem like swapping one human-driven pressure on evolution for another, but hopefully the impetus will be as positive as is the intention behind it.

If you still baulk at the idea of genetically modifying 'pristine' wild animals, then consider this. We have been genetically modifying wild animals for tens of thousands of years. Remember Higgs, my faithful GM wolf? All domestic species are genetically modified versions of their wild counterparts, but they have been altered by selective breeding rather than high-tech molecular methods. Furthermore, as we career into the Anthropocene, the epoch where human activity is dominating the globe, we influence the DNA of all living things, near and far. There may well be some ice-dwelling Antarctic microbes that are blissfully immune to the Anthropocene's onset, but as the world warms and the ice melts, this innocence will not last. Humanity's actions are now so far-reaching that they have become a global source of selective pressure. Evolution is speeding up as life responds to these environmental challenges. You could even go so far as to say that humans are genetically modifying *all* life on Earth, albeit inadvertently. There are no 'pristine' species. All life is touched in some form by the fingerprints of humanity.

We have developed an over-reliance on domestic livestock, and an under-appreciation of wild species. I think that, as humans became more dependent on domestic species and less reliant on wild ones, it fuelled a dichotomy between humans and the rest of the animal kingdom. Domestic creatures became commodities that could be traded, moved and manipulated. Wild species became resources that could be variously ignored, plundered and pushed around. In Victorian times, people thought themselves superior to the brutish and ungodly creatures of the natural world. It's a mind-set that persists to the present day and it fuels the laissez-faire neglect of the natural world. It makes it easier for us to ignore the extremes of selective breeding and the diminishing effects it has on the gene pools of domestic stock. It makes it easier for us to hide farm animals in industrial feedlots, with all of the associated welfare issues, and easier to treat what remains of the natural world as a resource that can be raided with impunity.

We like to think that we're set apart from the natural world, when in reality all living things are just twigs on the same tangled evolutionary tree. We breathe the same air and share the same planet. The natural world is vital to our existence and yet, because of our actions, it finds itself in danger. Domestic species have been the focus of our attention for so long, but now domestic *and* wild species need our help. Like it or not, we have become curators of the planet that we have come to dominate. It's our planet and it's our responsibility. Life is always changing, but with humans at the helm – informed humans, guided by science, fuelled by evidence, motivated by a desire to protect the only planet we can live on – we can help life to change for the better.

Additional Reading

This is not an exhaustive bibliography, rather a few selected sources that I particularly enjoyed.

First stop; visit the Centre for PostNatural History in Pittsburgh. This excellent museum is dedicated to all of the organisms that humans have intentionally altered. It is curated by Richard Pell and features Freckles the Spider-Goat and the gonads of poor old Jimmy Cat Carter: www.postnatural.org/

Chapter 1: The Wolf That Rolled Over

This well-written book explores the impact of ten different species that were domesticated by humans: Roberts, A. 2017. *Tamed: Ten Species that Changed Our World*. Penguin Random House, London.

In 1959 Russian geneticist Dimitri Belyaev set up an amazing experiment in the wilds of Siberia when he began to domesticate the silver fox: Dugatkin, L., A. and Trut, L. 2017. *How to Tame a Fox (and Build a Dog): Visionary Scientists and a Siberian Tale of Jump-Started Evolution*. The University of Chicago Press, Chicago.

Is this one of the oldest dogs in the world? Mietje Germonpré's analysis of the controversial Goyet skull: www.sciencedirect.com/science/article/pii/S0305440308002380

Why do so many tame animals have curly tails and floppy ears? Here is one possible explanation: www.genetics.org/content/genetics/197/3/795.full.pdf

Chapter 2: Strategic Moos and Golden Gnus

This is one of my favourite books. Artist Katrina van Grouw presents a carefully researched and beautifully illustrated history of selective breeding: van Grouw, K. 2018. *Unnatural Selection.* Princeton University Press, Woodstock.

How aurochs became cattle and colonized the world: onlinelibrary.wiley.com/doi/pdf/10.1002/evan.20267

In South Africa, hunters can pay to shoot unusually coloured wild animals that have been selectively bred for this purpose: theconversation.com/conservation-versus-profit-south-africas-unique-game-offer-a-sobering-lesson-82029

Are cattle, sheep and goats endangered species? www.ncbi.nlm.nih.gov/pubmed/17927711

The Taurus Programme seeks to bring back the aurochs via selective breeding: www.rewildingeurope.com/tauros-programme/

Chapter 3: Super Salmon and Spider-Goats

Salmon containing the genes of two other fish species is on sale in Canada: aquabounty.com/our-salmon/

The red canary was the world's first human-created transgenic animal. A must read for anyone interested in its story: Birkhead, T. 2003. *The Red Canary.* Bloomsbury, London.

In the 1950s, atomic gardening was all the rage: pruned.blogspot.com/2011/04/atomic-gardens.html

In 1982, scientists discovered how to make a mega mouse... and paved the way for others to deliberately modify the genomes of living things: www.nature.com/articles/300611a0

If you live in the USA and would like to buy a transgenic tropical fish, you can order them online: www.glofish.com/

The world's first CRISPR edited sheep were born on a farm in Uruguay. Learn about the science that produced them: journals.plos.org/plosone/article?id=10.1371/journal.pone.0136690

Chapter 4: Game of Clones

Find out more about the life and times of Dolly the Sheep: www.ed.ac.uk/roslin/about/dolly

A brief history of horse cloning: www.publish.csiro.au/RD/RD17374

Think carefully if you're considering having an animal cloned. These companies offer cloning services: www.crestviewgenetics.com, www.kheiron-biotech.com, www.viagen.com, en.sooam.com, www.boyalifegroup.com.

Perhaps the greatest legacy of Dolly is the creation of ethical stem cells, thanks to Shinya Yamanaka: www.cell.com/cell/fulltext/S0092-8674(06)00976-7

Chapter 5: Screwworms and Suicide Possums

The Royal Society published this useful booklet about gene drive: royalsociety.org/~/media/policy/Publications/2018/08-11-18-gene-drive-statement.pdf

Kevin Esvelt was the first person to propose a gene drive made from CRISPR. There is lots of useful information at his website: www.sculptingevolution.org/kevin-m-esvelt.

This is the paper where Kevin first proposes the CRISPR gene drive: elifesciences.org/articles/03401

Target Malaria wish to test their anti-malarial gene drive mosquitoes in parts of Africa. Here is their website: targetmalaria.org/

Chapter 6: The New Ark

On the scale of planetary change during the Anthropocene: Lewis, S., L. and Maslin, M., A. 2018. *The Human Planet: How We Created the Anthropocene*. Penguin Random House, UK.

Large animals are disappearing and humans are to blame. Felisa A. Smith's 2018 study: science.sciencemag.org/content/360/6386/310

Billions of populations of animals have been lost in recent decades, prompting scientists to warn of a 'biological annihilation.' A hard-hitting 2017 study from Gerardo Ceballos and colleagues: www.pnas.org/content/114/30/E6089

Humanity has wiped out 60% of animal populations since 1970, according to the 2018 Living Planet Report: www.wwf.org.uk/updates/living-planet-report-2018

In millions of years, future geologists will look back at the rocks that are being formed now and see a preponderance of chicken bones. Are we living in the era of the chicken? royalsocietypublishing.org/doi/pdf/10.1098/rsos.180325

This eloquent book spells out the link between industrial farming and biodiversity loss. A must read. Lymbery, P. 2017. *Dead Zone: Where the Wild Things Were*. Bloomsbury, London.

Chapter 7: Sea Monkeys and Pizzly Bears

'Just Add Water: The Story of the The Amazing Sea-Monkeys™' is a short film from the Centre for PostNatural History: postnatural.org/Just-Add-Water-The-Story-of-The-Amazing-Sea-Monkeys

Although biodiversity is taking a pummelling, some species are actually doing quite well. Read Thomas, C. 2017. *Inheritors of the Earth: How Nature is Thriving in an Age of Extinction*. Allen Lane, Penguin Random House, UK.

What's going on between grizzly and polar bears? They must have been interbreeding for a while: pgl.soe.ucsc.edu/cahill18.pdf

The Anthropocene could actually raise biological diversity... a thought-provoking commentary by Chris Thomas: www.nature.com/news/the-anthropocene-could-raise-biological-diversity-1.13863

Chapter 8: Darwin's Moth

This is Katherine Byrne's paper on the London underground mosquitoes: www.nature.com/articles/6884120

The mutation that turned the peppered moth black: www.nature.com/articles/nature17951

Have New York's white-footed mice evolved the ability to digest pizza and peanuts? www.ncbi.nlm.nih.gov/pubmed/28980357

City living is changing the wing shape of American cliff swallows: www.ncbi.nlm.nih.gov/pubmed/23518051

Intense trophy hunting is causing bighorn rams to evolve smaller horns: www.ualberta.ca/science/science-news/2016/january/the-measure-of-a-ram

Cities are causing life to evolve. Schilthuizen, M. 2018. *Darwin Comes to Town: How the Urban Jungle Drives Evolution*. Quercus, London.

'Unnatural Selection' by Adam Hart. This excellent Radio 4 programme reveals how humanity is altering the evolutionary paths of other creatures. www.bbc.co.uk/programmes/b06ztq58

Chapter 9: Resilient Reefs

Coral bleaching is occurring more frequently: science.sciencemag.org/content/359/6371/80

In 2015, the International Society for Reef Studies put out a consensus statement on climate change and coral bleaching. It's not good: coralreefs.org/wp-content/uploads/2014/03/ISRS-Consensus-Statement-on-Coral-Bleaching-Climate-Change-FINAL-14Oct2015-HR.pdf

Fancy learning how to make coral spawn in a tank? This is Jamie Cragg's 'how to' guide: www.ncbi.nlm.nih.gov/pmc/articles/PMC5743687/

This has to be one of the best workplaces in the world…. the Hawai'i Institute of Marine Biology where brilliant people are hatching brilliant plans to save the world's coral reefs: www.himb.hawaii.edu/

New interventions are needed to save coral reefs … A 2017 paper explains why this all matters: www.ncbi.nlm.nih.gov/pubmed/29185526

Ruth Gates and Madeleine van Oppen lay out the case for assisted evolution: www.pnas.org/content/112/8/2307

Chapter 10: Love Island

Find out more about the pioneering work of the New Zealand Department of Conservation's Kākāpō Recovery Team: www.doc.govt.nz/kakapo-recovery

This is a clip from one of the finest and funniest wildlife documentaries ever made; the BBC's 'Last Chance to See' featuring Mark Cawardine and Stephen Fry. Please watch 'Shagged by a Rare Parrot': www.youtube.com/watch?v=9T1vfsHYiKY

Listen to the Kākāpō Files, a podcast by Alison Balance: www.radionz.co.nz/programmes/kakapo-files

This excellent tableau details all of the birds and their family trees: public.tableau.com/profile/jonni.walker#!/vizhome/TheKakapo/Dashboard1

Chapter 11: Return to Wild

The story of the rewilding of Knepp Estate is both heart-warming and uplifting. This beautifully written book tells it all. Tree, I. 2018. *Wilding: the Return of Nature to a British Farm*. Picador, London.

For more on rewilding, read this. Monbiot, G. 2013. *Feral: Rewilding the Land, Sea and Human Life*. Penguin Books, London

Find out more about Rewilding Europe: rewildingeurope.com/ and Rewilding Britain: www.rewildingbritain.org.uk/

Find out more about de-extinction in my first book. Pilcher, H. 2016. *Bring Back the King: the New Science of De-extinction*. Bloomsbury, London.

Acknowledgements

I owe a huge debt of gratitude to the many people who have helped me while I have been writing this book. So I'd like to say thank you to:

Team Bloomsbury. Jim Martin, Anna MacDiarmid, Julia Mitchell and Kealey Rigden you are, as always, a pleasure to work with. Thank you for the support, the lunches and the excellent Christmas parties. I don't get out much, so these have been a highlight.

My talented illustrator, Amy Agoston. You are gifted and smart. Thank you for all the hard work you have put into bringing my book to life. I adore your artwork, and I adore you. I know you will do great things.

All of the people who have taken the time to chat with me, email me, donate photos and help me by proofreading drafts of the emerging manuscript. You are a brilliant bunch. This is a better book because of you. Thank you to Michael Kinnison, Ronald Goderie, Chris Thomas, Luke Alphey, Madeleine van Oppen, Alejo Menchaca, Jim Reynolds, John Ewen, Greger Larson, Mietje Germonpré, Francesca Dooley, Jeff Craig, Brenna Hassett, Juliane Kaminski, Anna Kukekova, Love Dalen, George Seidel, Randy Lewis, Gabriel Vichera, Andréas Gambini, Minda Davies-Morel, Katrin Hinrichs, Flavio Forabosco, Marc Maserati, Adrian Mutto, Melain Rodriguez, Pierre Taberlet, Marcel Niekus, Hermann Swalve, Guy Green,

Adam Hart, Austin Burt, Kevin Esvelt, Peter Dearden, Gary Lewis, Geoff Walker, Lluis Montoliu, Martina Crispo, Eric Hallerman, Dave Conley, Kris Huson, Zhiyuan Gong, Ross Barnett, Jan Zalasiewicz, Philip Lymbery, Max Holmes, Richard Harrington, Steve Foster, Andrew Hendry, Scott Hackett, Martin Phillip, Alec Kolaj, Phillip John Robbins, Lee Taylor-Wheal, Robert Brooker, George Perry, Ken Thompson, Quenton Tuckett, Ruth Gates, Bruce Robertson, Daryl Eason, Jason Howard, Jens-Christian Svenning, Ole Sommer Bach, Alastair Driver, Molly Merrow, Tim Birkhead, Bruce Whitelaw, Sergey Zimov, Nikita Zimova and Alli Cartwright.

Extra special thanks go to: Richard Pell, Director of the Center for PostNatural History, who took me on a virtual tour of his brilliant museum. One day, I promise to visit in person. Jamie Craggs and Keri O'Neil, who hosted my visit to the Horniman Museum and put up with me firing dozens of annoying questions at them while they were trying to get on with their work. Emma Hills, Jim Clubb and Jamie Clubb, who welcomed me to Heythrop Zoological Gardens and let me spend time with Glacier the silver fox. Charlie Burrell and Isabella Tree, who made me tea and took me on a guided safari tour of the Knepp Estate. I will return to find the elusive Purple Emperor. Geoff Martin from London's Natural History Museum, who found my Great-uncle Rick's peppered moth. Thank you. It meant so much. Andrew Digby, lead scientist with the Kākāpō Recovery Programme, who has fielded my repetitive Skype calls with good humour and bonhomie. Jane Bennett for her friendship and proofreading skills. Jess and Paul Semple, for fielding random questions about cattle penises and running such a top-notch farm full of wildlife. My friends Timandra Harkness, Jo Brodie, Tracey Mafe, Rachel Waters, Claire Wragg, Alex Cooper, Aby Hawker, Andrea Warrener-Grey, Justine Mallard and family, the Harrington family, Brian and Claire Dale, and Milly. Thank you for the doggy daycare, the childcare, the friendship, laughter and cups of tea. Love you all.

To anyone else that I've forgotten: sorry. You were amazing. I will buy you a pint next time we meet.

And finally: To my husband, Joe. Thank you for your editorial skills, endless patience and unwavering support. Thank you for always picking up the slack when I'm busy. You are always there when I need you. I love you and I really couldn't have done this without you. Now can we go to the pub?

To the rest of my family … to Amy, Jess and Sam. To my mum, Nijole, and to my genetically modified wolf, Higgs. Thank you for your love and support. Aš myliu Jus.

Index